Springer Geography

The Springer Geography series seeks to publish a broad portfolio of scientific books, aiming at researchers, students, and everyone interested in geographical research.

The series includes peer-reviewed monographs, edited volumes, textbooks, and conference proceedings. It covers the major topics in geography and geographical sciences including, but not limited to; Economic Geography, Landscape and Urban Planning, Urban Geography, Physical Geography and Environmental Geography.

Springer Geography—now indexed in Scopus

More information about this series at http://www.springer.com/series/10180

Malte Steinbrink · Hannah Niedenführ

Africa on the Move

Migration, Translocal Livelihoods and Rural Development in Sub-Saharan Africa

Springer

Malte Steinbrink
Geography
University of Passau
Passau, Bayern, Germany

Hannah Niedenführ
Institute for Migration Research
and Intercultural Studies
University of Osnabrück
Osnabrück, Niedersachsen, Germany

Translated by Shelby Long

ISSN 2194-315X ISSN 2194-3168 (electronic)
Springer Geography
ISBN 978-3-030-22840-8 ISBN 978-3-030-22841-5 (eBook)
https://doi.org/10.1007/978-3-030-22841-5

This Springer imprint is published by the registered company Springer Nature Switzerland AG
The registered company address is: Gewerbestrasse 11, 6330 Cham, Switzerland

Acknowledgements

Our gratitude goes first to *Shelby Long*, University of Osnabrück, for her concise interpretation of this book. Thank you Shelby, for your patience, flexibility, and friendliness—it was a joy to work with you! We would like to thank the team of the project "Towards Inclusive and Sustainable Rural Transformation in Sub-Sahara Africa" at the Centre for Rural Development (SLE) at Humboldt University Berlin, and especially *Dr. Susanne Neubert* and *Dr. Gabriele Beckmann* for their helpful suggestions. We would also like to thank *Erwin Vogl*, Geography Department of the University of Passau, for his professional support in creating the figures. In addition, we would like to thank our editor at Springer Nature, *Prof. Dr. Michael Leuchner*, as well as Mr. *Rajan Muthu*, for their support in the process of publishing. Also, many thanks to *Prof. Dr. Chris Rogerson*, University of Johannesburg, for encouraging us to write this book.

Finally, many thanks to our families, partners, and friends for their local and translocal "informal support" that made this book possible—*A strong network truly helps!*

Contents

About the Authors

Prof. Dr. phil. habil. Malte Steinbrink holds the *Chair of Anthropogeography* at the Faculty of Philosophy, University of Passau and is a member of the *Institute for Migration Research and Intercultural Studies (IMIS)* at the University of Osnabrück, Germany. He is also *Senior Research Fellow an der School of Tourism and Hospitality der University of Johannesburg (UJ)* in South Africa.

He researches and publishes in the fields of development and mobility research with a special focus on social inequalities, rural transformations, and urban development processes in countries of the Global South. The Social and Cultural Geographer is also known for his extensive contributions to the study of slum tourism (in particular the book *Slum tourism: Poverty Power and Ethics [Routledge, 2012]*). Geographical science studies and the potentials of social network analysis are further areas of focus in his research. Southern Africa is one regional focus of his empirical work.

Hannah Niedenführ is a member of the Institute for Migration Research and Intercultural Studies (IMIS) at the University of Osnabrück and the recipient of a scholarship from the Friedrich-Ebert-Stiftung. She studied African Studies, Law and Migration Studies at the University of Hamburg, INALCO Paris and the University of Osnabrück.

She is currently pursuing her doctorate with a dissertation on child migration in Burkina Faso. Her research also focuses on development processes and policies in the Global South, negotiations of migration and forced migration.

Chapter 1
Introduction

Abstract Africa is immensely influenced by the spatial movement of its population. *Africa on the Move*, however, is not about "African refugees" nor African migration to Europe. It is about much larger movements: the "everyday migrations" *within* Africa. And it is about the enormous relevance of migration to livelihood security on the African continent. *Africa on the Move* also makes clear that Africa is changing. Its social, economic, and ecological structures are subject to rapid processes of transformation. How do people organize their daily lives, either to improve their living conditions or to respond to these various changes? Here, migration plays as decisive a role as the social networks that span across vast distances between regions of origin and arrival. In particular, rural and urban areas in Sub-Saharan Africa are closely linked through informal relations. Circular migration and an intensive exchange of information, capital, and goods characterize the interdependence of rural and urban areas. These linkages are largely constituted in the informal structures of social networks which span territorial borders and, in turn, have a considerable impact on the everyday life on the African continent. Within many people's livelihood systems, the *Here* and *There*—e.g., in the city and in the countryside—have certain, partially complementary functions. It is only through a specific *In-Betweenness* that these can be linked and combined to contribute to livelihood security. This book is a plea for the *translocal dimension of development* to be given more attention than it has received so far. It is time both for development research and for practical development planning to adopt a *translocal perspective*. This introduction describes the basic idea of the book and explains its inner structure.

M. Steinbrink and H. Niedenführ, *Africa on the Move*, Springer Geography,
https://doi.org/10.1007/978-3-030-22841-5_1

Keywords Translocality · Migration · Translocal development · Rural structural change

Africa is immensely influenced by the spatial movement of its population. And since the "refugee debate" and the associated "European crisis", processes of African migration have received much more attention from the European public than they did just a few years ago. However, *Africa on the Move* is not about the "*rush on the European fortress*" and "*Africa's flight to Europe.*"[1] Instead, it is about much larger migration movements: the "everyday migrations" within Africa. And it is about the enormous relevance of migration for everyday livelihood security on the African continent.

Africa on the Move also makes clear that Africa is changing. Social, economic, and ecological structures are shifting as they are subjected to rapid processes of transformation. How do people organize their daily lives, either to improve their living conditions or to respond to these very serious changes? Here, migration plays as decisive a role as the social networks that span across vast distances between regions of origin and arrival.

Rural and urban areas in Sub-Saharan Africa are closely linked through informal relations. Circular migration and an intensive exchange of information, capital, and goods characterize the interdependence of rural and urban areas.[2] These rural–urban linkages are largely constituted in the informal structures of social networks which span space and borders and, in turn, have a considerable impact on the everyday life of a large proportion of the population on the African continent.

Within many people's livelihood systems, the *Here* and *There*—in the city and in the countryside—have certain, partially complementary functions. It is only through a specific *In-Betweenness* that these can be combined with each other and linked across space to contribute to the livelihood security.

The translocalization of livelihoods is not a new process but an increasingly important one. It will more strongly influence both urban and rural development in Africa in the future.

The emergence of *translocal livelihood systems* is not a new process but an increasingly important one. And this form of translocality is having an ever-greater impact on local conditions and developments. In the future, translocal livelihood systems will have an even greater influence on

[1]This is how Karin Finkenzeller titled an article in the German magazine *Welt am Sonntag* (October 9, 2005, p. 9).

[2]"Migration in Africa is predominantly circular in nature and characterized by an emigration-diaspora-return continuum or origin-destination continuum." (UN 2018: 58)

both urban development dynamics and processes of rural structural change in Africa than they already do today.

Translocal processes that take place outside the formal markets for labor, capital, and goods and beyond welfare state interventions constitute an essential element of the distribution of wealth on various spatial scales. Hence, taking these informal flows and interdependencies into account offers significant explanatory potential for the investigation of unequal living conditions in different spatial contexts, for it enables us to analytically comprehend certain dynamics of the emergence of spatial and socio-structural disparities.

Although complex rural–urban interdependencies have increasingly gained recognition in the development debate since the late 1980s, binary spatial-analytical categories continue to dominate empirical research and development practice. This binary categorization fosters a compartmentalized perspective: Town and country are regarded as largely separate spheres. The disadvantage of such a view lies in the danger of too easily getting stuck in the "container bias" of conventional spatial concepts, remaining caught in a "*territorial trap*" (see Sect. 2.1.2).

In any case, the fact that more and more people are organizing their livelihoods translocally seems to present great challenges not only to science but also, and above all, to politics, planning, and development cooperation.

That "development experts"—whether scientists or practitioners in the field—do not pay enough attention to informal rural–urban interlinkages—despite a general awareness of their evident empirical basis and relevance—is presumably due mainly to institutional factors. Conventional rural/urban thinking has led to the establishment of two specialized expert communities, one of which is concerned with urban development and the other with rural development in the Global South. These two circles of development experts have often remained somehow alien and acted largely independently of each other. In this respect, the call for a stronger consideration of translocal urban–rural relations in practice is ridden with prerequisites. This is especially true as experts on both sides see a need for intervention not in the management of rural–urban structures but—depending on their own professional provenance—with regard to the problems of either urban or rural development. In addition, the administrative logic of planning and politics continues to focus on territorial units. Even in regions with strong emigration or immigration, politics often adheres to sedentary or local development strategies; migration and development are still regarded more as alternative rather than interdependent processes.

Despite these institutional obstacles, it is time both for development research and for practical development planning to adopt a *translocal perspective*. We should acknowledge that in many cases the rural and the urban are socioeconomically intertwined to such an extent that a separating view of *Here* and *There* hardly makes sense anymore. It simply does not help us understand how people in African cities and villages organize their daily lives, much less the complexity of rural–urban relations and their effects. Instead, it is necessary to recognize that mobility and spatial interconnectedness are normal aspects of everyday life for a large part of Africa's population and should thus be seen as integral factors of African development.

It is time for development research and practical development cooperation to take on a translocal perspective.

This book is therefore a plea for the translocal dimension of development to be given more attention than it has received in policy-making so far. If development policy is to become more effective and come closer to its high goal of "sustainable, socially inclusive development," developing a sensitivity for the dynamics in translocal social contexts and their potency is imperative. This also means adapting conventional policy and planning approaches by recognizing translocal networks and translocal flows as defining forces of social, economic, and ecological change.

The circumstances of this book's creation are ultimately an expression of exactly that new consciousness for which we advocate. It was written in the context of a project funded by the *German Federal Ministry for Economic Cooperation and Development (BMZ)* and conducted by Dr. Susanne Neubert at the *Centre for Rural Development (SLE)* at Humboldt University Berlin. Within the framework of this application-oriented project, whose aim was to analyze rural structural change in Africa and identify strategies and measures to make it more socially inclusive and sustainable (see Rauch et al. 2016: ii), we were asked to write a report on the significance of translocal livelihoods for this process. This report (Steinbrink and Niedenführ 2017) served as the basis for this book.[3]

[3]We would like to thank the team of the project "Towards Inclusive and Sustainable Rural Transformation in Sub-Sahara Africa" (Engel et al. 2017) and especially Susanne Neubert and Gabriele Beckmann for their helpful suggestions.

Since our focus lies primarily on rural areas in Sub-Saharan Africa, the present exposition is also caught in the old rural/urban dualism to a certain extent. The decisive difference, however, is the *translocal perspective*: we examine central aspects and processes of rural structural change against the background of translocal realities and discuss the significance and influence of translocal livelihoods on rural development.

Since the translocality approach is still relatively new to both science and development cooperation, the theoretical foundations are given ample space in this book: Chaps. 2–4 provide the necessary conceptual tools. These chapters are based largely on preliminary work by the first author (Malte Steinbrink) and form the theoretical basis for the explanations in Chaps. 5 and 6, in which the patterns, rationalities, and effects of translocal livelihoods in Sub-Saharan Africa are presented. The aim of these two chapters is to give an overview of the research landscape.

To begin with, Chap. 2 explains the need for a shift in perspective by highlighting certain conceptual pitfalls of conventional approaches in migration studies as well as in development research and practice. An outline of a translocal perspective is then presented as an alternative concept to conventional container-spatial thinking.

Chapter 3 addresses the concept of translocal livelihoods and discusses the central concepts (e.g., *translocal household* and *translocal community*) and methodological challenges of translocalizing livelihood and vulnerability research based on the current state of the debate.

Chapter 4 fleshes out the concept of translocal livelihoods in order to establish some theoretical foundations for answering the question of how the translocality of livelihoods in Africa is influenced by different degrees of vulnerability and why the translocalization of livelihood systems occurs. We present an analytical model of translocal livelihoods which focuses on the dialectic relationship between agency and structure on the one hand, and the specificities of agency in the context of vulnerability on the other. At the center of this model are the strategies of translocal *economic diversification*, *transmigration* and the importance of translocal *social networks*.

Chapter 5 addresses the phenomenon of translocal livelihoods in Sub-Saharan Africa. After a brief description of the particularities of the African urbanization process (Sect. 5.1), we attempt to quantitatively assess the significance of translocal livelihoods in Sub-Saharan Africa (Sect. 5.2). Subsequently, we discuss spatial and temporal patterns of translocalization (Sect. 5.3) and the social structuring associated with it, e.g., with regard to age, gender and education level. This chapter also outlines a "phase model of translocalization" (Sect. 5.3.3) and deals with changes in translocal communication patterns (Sect. 5.3.5).

Chapter 6 analyzes the effects of translocal livelihood organization on different aspects of rural structural change. Economic, ecological, and social dimensions are considered. Section 6.1 ("*Economic dimensions*") first focuses on the scope and significance of remittances. The following chapters deal with the distribution of workforce and workload in translocal contexts (Sect. 6.1.2) and the importance of translocal networks for knowledge transfer and innovation processes in rural areas (Sect. 6.1.3). Section 6.2 examines the relationship between translocal livelihoods and changes in natural environmental conditions. Here two questions are addressed: How do environmental changes affect translocalization processes? And to what extent does the translocality of livelihoods influence the use and management of natural resources? With regard to the social dimension, Sect. 6.3 looks in particular at education, gender relations and health in translocal contexts.

Finally, Chap. 7 offers a summary of the effects of translocality. Here we attempt to provide more general answers to the questions of (a) if translocal livelihoods have a balancing effect on the patterns of social and spatial inequalities, (b) if the translocality of the livelihoods should be regarded as a temporary phenomenon and (c) to what extent translocality can contribute to solving the poverty problem in Sub-Saharan Africa.

References

Engel E, Rettberg S, Rauch T, Neubert S, Richter D, Minah M, Berg C (2017) Towards inclusive and sustainable rural transformation in Sub-Sahara Africa. Selbstverlag Humboldt-Universität zu Berlin, Berlin

Rauch T, Beckmann G, Neubert S, Rettberg S (2016) Ländlicher Strukturwandel in Subsahara Afrika. SLE Discussion paper 01/2016. Available via SLE. https://www.sle-berlin.de/files/SLE%20Discussion%20Paper%2001-Strukturwandel%20mit%20Cover.pdf. Accessed 18 Mar 2019

Steinbrink M, Niedenführ H (2017) Afrika in Bewegung: Translokale Livelihoods und ländliche Entwicklung in Subsahara-Afrika. Transcript, Bielefeld

UN (2018) Economic development in Africa report. Migration for structural transformation. New York, Geneva

Chapter 2
Time for a Rethink

Abstract The translocal perspective is not yet fully established in social science. As an analytical perspective, translocality represents an alternative to some dominant approaches. In this chapter, we demonstrate the need for a change in perspective by highlighting the conceptual pitfalls of conventional approaches in migration studies and development research and practice. First, we take a critical look at previous research on the relationship between migration and development (Sect. 2.1), exploring how interdisciplinary research on this nexus has long been caught in two methodological traps: an "ideological trap" and a "territorial trap". We then examine conventional urban/rural thinking in development research and cooperation (Sect. 2.2), showing that the container-spatial bias of this dichotomizing view can lead to some grave misconceptions of rural and urban development. From the analytical limitations of the traditional views, we derive the need for a rethink and introduce the idea of translocal livelihoods as a conceptual link between migration research and development studies (Sect. 2.3).

Keywords Translocal perspective · Container-spatial thinking · Rural/urban thinking · Territorial trap · Ideological trap · Development theory · Migration theory

The translocal perspective is still relatively new.

The translocal perspective is relatively new and not yet fully established in social science. As a research perspective, translocality represents an alternative to some dominant approaches. In the following, our aim is to demonstrate the need for a change in perspective by highlighting the conceptual pitfalls of conventional approaches in migration studies and development research and practice.

M. Steinbrink and H. Niedenführ, *Africa on the Move*, Springer Geography,
https://doi.org/10.1007/978-3-030-22841-5_2

First, we take a critical look at previous research on the relationship between migration and development (Sect. 2.1); then we examine conventional urban/rural thinking in development research and cooperation (Sect. 2.2). From the limitations of the traditional views, we then derive the need for a rethink (Sect. 2.3).

2.1 Research on Migration and Development: Caught in Two Traps

At the end of the 1990s, McDowell and De Haan (1998) stated that development research had long insufficiently concerned itself with migration-related issues. This often-quoted statement is no longer valid today, nor could it be accepted without reservation even at that time. For example, Geiger and Steinbrink (2012) show for the scientific landscape in Germany that the focus of migration-related research has by no means been restricted to the German or European context. Indeed, as early as the late nineteenth and early twentieth century, the focus lay clearly on those world regions labeled "underdeveloped" by Western industrial nations after the Second World War. Early migration research was mainly concerned with internal migration processes; nomadism, "rural exodus", politically induced flight, expulsion, resettlement, urbanization, and labor migration were described. This focus on smallscale, often circular migration processes in colonial regions and in the later decolonized states persisted after 1945 (see Geiger 2011: 35–41). Thus, those traditional themes of early German migration research in "foreign, faraway lands" were conveyed first into "research in developing countries" after the Second World War and later into development studies, which began gaining ground in the 1970s.[1]

Given this long tradition of research on migration in the "Global South", German-speaking scholars can be absolved of neglecting the topic of "migration and development". On the contrary, German migration research began there and only turned its attention to the German and European context much later. More revealing than the question of *whether* earlier

[1]On the history of development research in Germany, see Scholz (2004) and Bohle (2007).

research dealt with migration-related aspects of development is, therefore, the question of *how* it addressed these questions.[2]

In the following, we will briefly show that most interdisciplinary research on the nexus of migration and development has been caught in two methodological traps: an "*ideological trap*" and a "*territorial trap*".[3]

2.1.1 The Ideological Trap

It is a truism that scientific research is always embedded in the broader context of society. The orientation of research on migration in countries of the South has always manifested the respective (political-ideological) zeitgeist. As variable as the horizons of meaning of the amoeba-like term "development" itself are the academic answers to the ultimately far too general question of whether migration has a positive or negative effect on "development"; they have varied greatly according to the times (see Martin and Taylor 1996). Thus, De Haan (2000) states that, within international development research, there is no agreement on the role of migration in the development process. In fact, the assessments are contradictory (see Table 2.1): While "*migration optimists*" assume, along the lines of neoclassical equilibrium models, that migration leads to a reduction of interregional inequalities, "*migration pessimists*" argue that no such balancing tendencies exist, and that migration actually exacerbates the development gap (De Haas 2010). Even the images of migrants underlying the respective studies indicate the controversy between these contradictory viewpoints (see Table 2.1). The migration optimists regard the migrant as a rational actor who, as an optimizer in the sense of a *homo economicus*, can compare the different opportunities in the region of origin and arrival and weigh up the expected benefits in order to make a free and self-determined migration decision. The migration pessimists, on the other hand, tend to equate migrants with *victims*—be it of economic impoverishment and exploitation, political or social discrimination, or progressive environmental destruction.

---

[2]For a review of the migration-related studies of German-speaking development researchers, exemplified by research in Africa, see Wenzel (2012).

[3]For more comprehensive remarks on this, see Geiger and Steinbrink (2012: 16–25).

Table 2.1 Optimistic and pessimistic positions in the migration-development debate

	Migration optimists	Migration pessimists
Effects of migration on development	**Reduction of regional disparities** (→ Equilibrium) **Growth and prosperity everywhere**	**Intensification of regional disparities** (→ Inequality) **Impoverishment in the context of origin**
Image of migrants	**Proactive actors** – *"homo economicus"/ "optimizer"* – Freedom of decision and action – Oriented toward maximizing benefits (→ *"Maximization Rationality"*)	**Victims** – Expelled – Exploited – Compelled to act (→ "*Survival Rationality*")
Ideological alignment	**Modernization** – Faith in progress, growth, technology and markets – Incremental *"catch-up development*"	***Dependencia*** – Fundamental critique of Imperialism and Capitalism – *"Development of underdevelopment*"

Source Own illustration

These two opposing positions are also reflected in the standard repertoire of theoretical models and approaches that migration researchers have historically used in their analyses (see Table 2.1). What is more, simply in choosing their approach, scientists have implicitly aligned themselves with one of the two grand theories in the development theoretical dispute: *modernization* or *dependencia*.

In modernization theories: *Migration as a transitive phenomenon of "catch-up development"*.

The classical, economically oriented migration studies approaches conceived the movements of economic actors as chosen courses of action, preceded by decisions based on economic self-interest. Very prominent is the *Todaro model* (Todaro 1969), which considers migration decisions to be determined by wage differentials and employment rates. The other push–pull models are theoretical extensions of this idea and are ultimately based on the macroeconomic idea of factor mobility or on the dualistic *surplus labor model* by Lewis (1954). The approach of the *New Economics of Labour Migration* (Stark and Bloom 1985; Stark 1991) broadened this view of migratory phenomena by taking as its basic unit of analysis not the autonomous individual but households, and by addressing the action-guiding aspect of risk minimization. However, this approach remained bound to the individualistic-functionalist tradition of economic migration theories to the extent that it largely ignored noneconomic factors and structural frameworks, continuing to emphasize freedom of choice. Migration researchers who based their studies on these models ultimately viewed migration through a modernization-theoretical lens as a logical phenomenon in the course of incremental "catch-up development," i.e., a transitional phase of the industrialization and urbanization process (see Table 2.1).

In dependency theories: *Migration as a transmission belt in the "development of underdevelopment".*

Strongly inspired by the *Dependencia School*, the neo-Marxist and structural–theoretical approaches accused the modernization theorists of neglecting the structural causes of migration and of considering migration processes outside the historical-material context of world economic realities. For their part, they emphasized the political-economic aspects of migration phenomena and their entanglement in a historically rooted network of economic interconnectedness characterized by asymmetric dependencies.[4] In contrast to neoclassical migration theories, migration was not understood as an economically rational action resulting from a free-will decision but as a kind of transmission belt in the process of the "*development of underdevelopment*" (Frank 1969).

Table 2.1 summarizes the two fundamentally opposing positions in the debate about migration and development.

Conforming to these fundamentally dissimilar approaches, research from these two academic camps assessed the effects of migration on economic and social development processes very differently: While the modernization-inspired approaches to migration studies tended to predict growth and balance, the approaches critical of imperialism and capitalism predicted the continuation and intensification of poverty and inequality. Accordingly, one cannot discern any uniform line in the literature as to whether migration has a positive or

---

[4]Authors like Wolpe (e.g., 1972), McGee (e.g., 1982), Standing (1985) and Potter and Unwin (e.g., 1995) placed migratory movements in countries of the Global South in the context of postcolonial structures and capitalist transformation processes in peripheral capitalist states. These systemic or structurally oriented approaches attempted to explain migration in terms of the functional requirements of growth and change within the capitalist world system.

negative effect on development processes in general, or on the well-being of migrants and those left behind in particular; one can only distinguish fluctuations in which viewpoint has predominated at any given time.[5]

The scientific debate on the relationship between migration and development has thus taken place less on an empirical-analytical level than on a political-ideological one. This explains why the debate has ultimately failed to grasp the complexity of the issue. The major controversy between the two grand theories (modernization and dependencia) drove development-related migration research into a kind of "*ideological trap*".[6]

The far-reaching theoretical upheavals which began in the early 1990s also took hold in development research. After the collapse of the Soviet Union and the putative end of the battle between the political systems, not only did the division of the world into a "first", a "second" and a "third" world become obsolete, but the dispute between modernization and dependency was settled as well.[7] In turn, development research entered a phase of unsettled disorientation and fled conceptually into "*medium-range theories*" (Hein 1998). However, out of this emerged—if at first rather hesitantly—a creative search for new lines of theory (see Dörfler et al. 2003). From the mid-1990s onward, modified questions and partially new approaches took center stage in explaining migration phenomena and their development-related consequences. Migration-related development research drew on both the

---

[5]See, among others, the literature overviews of McDowell and De Haan (1998), as well as Waddington and Sabates-Wheeler (2003), de Haas (2010) and Steinbrink and Schmiz (2019).

[6]Here we refrain from comparing and weighing up the various approaches to migration theory in detail, but it should be emphasized that while each has valuable explanatory potential, none of them alone suffices to analytically grasp the complexity of the various migration phenomena. Although their basic assumptions, arguments, and conclusions are sometimes mutually exclusive, they are complementary in that, strictly speaking, they often attempt to explain different matters. Migration is simply too heterogeneous and context-specific in its manifestations to be regarded as a uniform object of research. Nevertheless, in the past, migration was largely treated as if it were *a single* phenomenon that could be explained by *a single* theory. For this reason, since the 1990s, Massey et al. (1993) and many other migration researchers have rightly argued that there cannot be *one* "grand theory" of migration studies.

[7]Menzel (1993) spoke in this context of a "failure of the grand theories," which had structured development discourse for about two decades.

discipline's own newer theoretical offerings and contemporary concepts from international migration studies (see Sect. 2.1.3), thus partly liberating itself from the ideological trap.

2.1.2 The Territorial Trap

No migration without demarcation

Mobility/migration is generally understood as physical movement between different localities. This reflects the conventional understanding of space, as well as the spatially bound and space-binding conceptualization of mobility/migration. Mobility (in the sense of commuting between different functional locations) is usually regarded as unproblematic; however, as soon as territorial political borders are crossed—and mobility takes place as (international) migration (in the sense of a long-term residential relocation)—it becomes problematized and politicized. Migration thus presupposes *qua definitionem* borders: No migration without borders! Due to this space- and border-bound definition, the research of migration almost inevitably reproduces a container-spatial perspective; and the border is continuously confirmed without itself being explicitly addressed or questioned.

In migration research, "space" is usually regarded as a given, static signifier that forms a seemingly natural frame of reference.

Accordingly, migration scholars' research activity has always been *spatially bound* and *space-reproducing*. Their main task has been to track, describe, analyze, and predict migration processes and their associated (spatial) consequences. In most studies on migration today, territorial space and the borders that constitute it are still regarded as a given and static reference, a category that seems to naturally frame the investigation of migration. This fundamental methodological problem of hardly being able to examine society or social phenomena, such as migration, independently of specific reified spatial constructs is called the "*territorial trap*" (Agnew 1994). The pitfalls of spatially moored thinking are indeed difficult to circumvent, and such thinking shapes migration research in general (see Berking 2006).

In the 1990s, social science began to recognize the territorial trap and the associated problem of "*methodological nationalism*" (Wimmer and Glick Schiller 2002), or *methodological territorialism*, and thus embarked on a search for alternative ways of thinking. In the course of this search, researchers embraced impulses from migration studies on the one hand and from development research on the other.

2.1.3 *Impulses from International Migration Studies and Development Research*

In international migration research, various approaches to migration studies have been established which can each be interpreted as an effort to conceptually overcoming a spatially bifurcating view of countries of origin and arrival.

Transnationalism: *Plurilocal social interconnections between countries of origin and arrival.*

Probably the most influential attempt at thinking beyond national containers comes from the US-American discussion of so-called *transnational social spaces* (see, e.g., Glick Schiller et al. 1997; Pries 2008; Faist 2000). Its starting point was the empirical observation of a supposedly new type of migration from the Caribbean states, Mexico and the Philippines—i.e., from so-called developing and newly industrializing countries—to the United States. Since the 1980s, circular patterns of migrants oscillating between countries of the origin and arrival have increasingly been observed. Viewed in light of *transnationalism*, migration is no longer seen as a single, unidirectional movement and final departure from the migrant's previous social context. Instead, constellations of multilocal social interconnectedness and plurilocal life realities that span across national boundaries are emerging.[8] This "new" form of migration does not involve a complete detachment from the societal context of origin and social integration in the country of arrival; instead it is expressed in a permanent "in-betweenness" and is thus often referred to as *transmigration*.[9]

Conceived at the beginning of the 1990s as an alternative to methodological nationalism, this *transnational* perspective has not only called for a rethinking of questions of social assimilation/integration in the arrival context; it has also concerned itself with the branch of migration research explicitly devoted to the effects of international migration on

---

[8]For an example of early European research in this field, see Müller-Mahn (1999). In an empirical case study of south–north migration based on the example of Egyptian "*sans papiers*" in France, the author examines the relationship between irregular immigration and irregular employment.

[9]"Transmigrants are immigrants whose daily lives depend on multiple and constant interconnections across […] borders and whose public identities are configured in relationship to more than one nation-state." (Glick Schiller et al. 1997: 121)

origin contexts. Accordingly, migration research has increasingly examined the development-related effects of transnational interconnectedness on transmigrants' origin countries. This research direction was established concurrently with the discourse on *Migration & Development* in international politics (Geiger and Steinbrink 2012; Steinbrink and Schmiz 2019). Today, the idea of transnationalism effectively forms the conceptual basis for political argumentations that postulate the positive impact of remittances, diaspora groups, remigration, brain gain/circulation, etc.—although, in contrast to the political rhetoric, the actual research results certainly do not paint an unequivocally positive (or development-conducive) picture.

Migration system *as a complex, multidimensional interrelationship between origin and arrival region.*

The *migration system approach* is another perspective for considering the origin and arrival regions in their systemic interaction instead of separately. This approach, applied by Kritz et al. (1992) to international migrations, notably originates in development-geographical work on rural–urban migrations in Africa (Mabogunje 1970). The migration process is interpreted here merely as one component of a complex fabric of exchange between the regions of origin and arrival, i.e., a larger systemic framework, which is also structured by economic, demographic, social, political, and legislative factors. It is assumed that migration systems arise through the perpetuation of migration-inducing, migration-intensifying or migration-influenced factors at different levels (micro to macro).

Social capital and social networks.

Conceptually closely linked with the perspective of transnationalism and the migration system approach are the different approaches to *social capital* and *social networks* in migration research. The concept of transnationalism in particular is hardly conceivable without thinking in terms of networks, for it is the social relationships/interactions between individual and collective actors that constitute the border-spanning social fields addressed in transnationalism research. Social networks also play a central role in the migration system approach. To a certain extent, they bridge the theoretical gap between micro- and macro-perspectives.[10] The migration studies analysis of this meso-level focuses predominantly on those social relationships that reduce the costs and risks of migration.

[10]Faist (2010) also speaks of a "meso-level theory".

Thus, the analysis of migration networks contributes significantly to an explanation of the dynamics in migration systems. In addition, it offers an important perspective for a better understanding of the phenomena of chain migration, the "*cumulative causation of migration*" (Massey et al. 1994) and the establishment, consolidation, and adjustment of migration paths and space-spanning social fields. Within network-related migration research, *social capital* has become a very prominent concept (Palloni et al. 2001).

The boom in network approaches in international migration research reflects a general trend in the social sciences, which has been apparent for around twenty years. In international development research, too, the network is enjoying increasing popularity as an analytical concept. After the supposed "failure of the grand theories" (see Sect. 2.1.1), development research turned primarily to more pragmatic actor- and action-oriented approaches. Indicative of this was the emergence of *vulnerability and livelihood research*,[11] which has advanced since the 1990s to become one of the most prominent perspectives of development research (see Scoones 2009).

The vulnerability concept interprets individuals and groups as actors in risk situations and focuses on the organization of livelihood systems. The risk constellation is thereby understood as a multiplex situation within which vulnerable individuals and groups (mostly households) develop adaptation, coping and livelihood strategies. The focus lies on resource bases and strategic action in the context of crises and uncertainty. In this research area, interpersonal relationships—i.e., social networks—are classified as an important part of the available *livelihood assets*; they are regarded as a social resource (*social capital*) that can be accessed in the event of a crisis.[12]

---

[11]See Chambers (1989); and on the adoption of the idea of vulnerability in German development research, see Bohle (1993) as well as Bohle and Watts (1993). These concepts have been used as a starting point and analytical framework for various extensive development-geographical studies.

[12]In fact, the "livelihoods perspective and vulnerability thinking have much in common (Bohle 2009). Both require a concise analysis of the baseline conditions and triggers of people's vulnerability to, for instance, natural hazards, economic shocks or political violence. Both sharpen the researchers' foci to asses the capacities and strategies of actors, households and communities, and their respective embeddedness in networks and their position in the broader society. And both bear in mind the structural formations of people's vulnerability or livelihood

However, livelihood studies' strong concentration on the microlevel of the actors and its container-spatial conceptualization of central analytical categories, e.g., household and community (see Sect. 3.2 for details) lead the microlevel to be equated with the local level. Supposedly, a kind of spatializing reflex leads researchers to often interpret the livelihood systems they analyze as locally bound. This critically abridged conclusion is another form of the territorial trap. Only recently have *translocally organized livelihood systems* been given greater emphasis (see Chap. 3 for a detailed description), which, in turn, has brought migrations and space-spanning social networks into focus as integral components of livelihood security. This means that the network perspective also conceptually links migration research with development research.

The concept of translocal livelihoods *is an action-centered approach that explains migrations and space-spanning networks in the context of risk and uncertainty.*

The *concept of translocal livelihoods* is a migration studies approach that originates in development research and attempts to explain migrations and space-spanning social networks in the context of risk and uncertainty. This idea expands livelihood research, allowing it to overcome the container-spatial bias of conventional development research and development cooperation. In this way, the concept offers an action-centered approach for analyzing various space-spanning forms of economic, migratory, and social interactions related to development issues, as well as a means to understand the significance of these translocal interactions for the livelihoods of *movers* and *stayers*.[13] In addition, the concept has considerable explanatory potential for the processes reproducing, equalizing, or intensifying social and regional disparities. In this respect, this relatively young approach can contribute in several ways to the current debate around the highly politicized nexus "*Migration & Development*" (see Fig. 2.1).

security in terms of social norms, values, legal rules, and governmental policies." (Etzold 2017: 46)

[13]See, among others, Lohnert and Steinbrink (2005), Thieme (2008), Zoomers and van Westen (2011), Steinbrink (2012), Steinbrink and Peth (2014) and Etzold (2017).

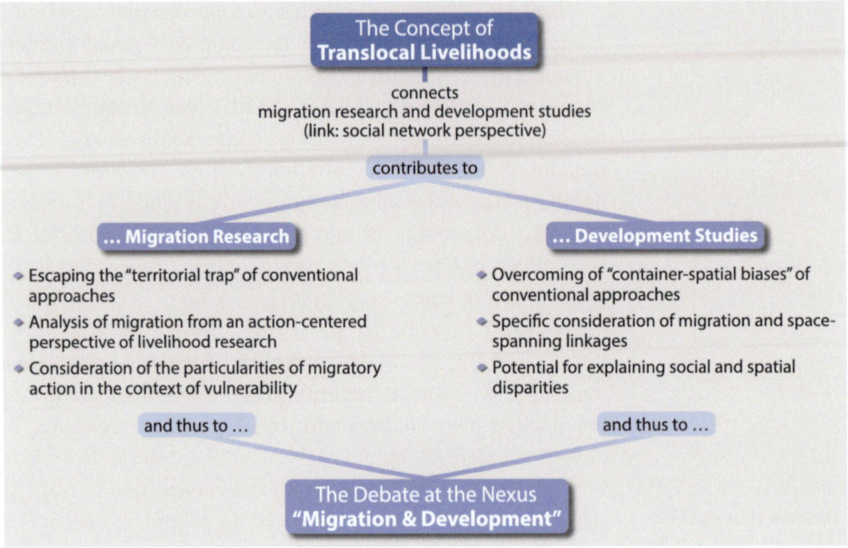

Fig. 2.1 Translocal livelihoods as a conceptual link between migration research and development studies. *Source* Own illustration

2.2 The Limits of Rural/Urban Thinking

Rural and Urban are categories for the spatialization of analytically dualistic relationships.

On a conceptual level, "the rural" and "the urban" are irreconcilable opposites. Both are conceivable only in contrast to the other. It could be pointedly argued that this antonymous linguistic relationship is a hidden core problem of development research, and of geographical development research in particular. Urban and rural are inherently spatial concepts, and as such, they are fundamentally constituted in the spatial categorization of other pairs of opposing concepts. Only through the contrariety of the other chosen pairs of terms does the rural/urban dichotomy emerge. The terms rural and urban are bursting with popular and scientific connotations that are so diverse and sometimes contradictory that the terms ultimately lose any significant analytical value. Nevertheless, this dichotomous rural/urban concept is deeply rooted in "development thinking" and leads to the separate consideration of "rural" and "urban" problems.

Dichotomous rural/urban thinking is deeply rooted in development research.

Considering the empirical reality that complex social interactions and interweavings link urban and rural areas in many countries of the Global South, this thinking is proving to be inadequate. That is why a central thesis of this book holds instead that the socioeconomic interconnectedness of rural and urban areas is often so strong that a separate consideration can lead to blatant misjudgments in the assessment of many specific problems.

In the following, the reasons for the astonishing tenacity of a dichotomizing rural/urban thinking in the development theory and policy debate will be examined: How have the debates on development theory and development policy at different times manifested themselves in rural/urban thinking and contributed to its reinforcement?

2.2.1 *Rural/Urban Thinking in Development Theory: Dualistic Spatializations*

In international development research, rural and urban have always been central analytical categories. Researchers' approaches and conclusions regarding the issues of urbanization, rural–urban migration and the structural relationship between rural and urban areas tend to reflect the zeitgeist of development theory and politics prevailing at the particular time. Each of the two major theoretical schools of thought has used the concepts rural and urban to *localize their respective central dualistic categories*. This spatialization of nonspatial categories and the dualistic logic of classical rural/urban thinking is illustrated in the following by means of a comparison (for a detailed account, see Steinbrink 2009a: 22–38).

2.2.1.1 Rural/Urban as Antithetical Dualism

Modernization-theoretical rural/urban thinking: *Antithetical dualism of tradition and modernity.*

According to one of the fundamental assumptions of modernization theories, the irreconcilable discrepancy between "traditionality" and "modernity" shapes the internal structure of so-called developing countries. The traditional and the modern are consequently regarded as an antithetical dualism.[14] At the same time, modernization theories assume this

[14]This model of dualism assumes that societies are divided into two incompatible sectors: the modern, dynamic sector integrated into the world market on one hand and the traditional, backward, and stagnating sector on the other. In the view of modernization theorists, both are subject to different developmental dynamics, each with its own rules (see Gilman 2018; Pieterse 2010).

dualism to be merely a transitional phase in the development process, believing that the premodern sector dissolves in the course of successive development progress ("catch-up development").

Considering the central supposition that developing countries would follow the "pattern" marked out by the historical European and Anglo-American macroeconomic experiences, it is unsurprising that the concepts of "rural" and "urban" lead a quiet spatialization of this dualism model; the key terms "traditionality" and "modernity" appear to translate one-to-one into spatial categories. Thus, the urban is perceived as the spatial carrier of all those attributes associated with "development" and as the site where all those modernization processes take place, which contribute to *Take-Off* and overcoming "underdevelopment". The rural, on the other hand, symbolizes economic, cultural, and social stagnation and is thus the spatial carrier of the traditionality and backwardness which are overcome at the end of the development process. Analogous to the traditional/modern dualism, the rural and the urban thus stand parallel as two isolated spheres, which exhibit different and relatively autonomous development dynamics. In light of the postulated evolutionary progression, they are regarded as quasi-phases in the development process. Social and economic development is equated with urban development. Accordingly, modernization theorist sees the rural and the urban, to some degree, as spatially manifesting the model of a dualistically structured society, which they understand as a transitional phase in the process of growth and development. In this way, the antithetical dualism of "rural/urban" seems to arise self-evidently from the antithetical dualism of "traditionality/modernity".

2.2.1.2 Rural/Urban as Dialectic Dualism

From the mid-1960s onward, the dependency-theoretical critique set at the roots of modernization theories. First and foremost, the basic assumption that underdevelopment has endogenous causes was radically questioned on a theoretical level. Instead, explanatory models began to focus on the entanglements of "underdeveloped countries" in a network of dependencies on industrial states, characterized by asymmetric relations (see Senghaas 1972, 1974; Evers and Wogau 1973). Within dependency theories, the concept of "structural heterogeneity" advanced by Córdova (1973) evolved into the central analytical approach.[15] According to this concept, structurally heterogeneous societies differ from structurally homogeneous ones in that different, specifically interacting modes of production coexist within them. In this respect, dependency theories' concept of structural heterogeneity serves a similar function as the dualism model in modernization theories: Both approaches conceptualize the inner structure of peripheral societies. Like the modernization-theoretical concept of dualism, the concept of structural heterogeneity assumes that different segments of society exist. However, in dependency theories, the distinction is not made between traditional and modern segments, but between segments of capitalist and noncapitalist or precapitalist modes of production. Thus, this concept also makes use of an analytical dualism ("capitalist/noncapitalist"); however, it is not understood as antithetical, but—because the interdependencies and the interplay between the sectors are the focus—as *complementary dualism.*

Dependency-Theoretical Rural/Urban Thinking: *Complementary dualism of "capitalist/ noncapitalist".*

The dependency-theoretical critique also referred to the dualistic rural/urban concept of modernization theories, though it did not aim at dismantling the "spatialization" of dualism itself. Rather the rural and the urban remained, even in this critical examination, the spatial carriers of the alternative political-economic concept's central categories. The concept of structural heterogeneity was translated into the "spatial" too, as rural and urban were used as categories for locating the different modes of production, or for locating production and externalized reproduction. The countryside is thus regarded as the site of the precapitalist or noncapitalist sector (subsistence economy), while the city is interpreted as the space of the dominant capitalist mode of production—and as the center of capital accumulation, which benefits above all from the externalization of reproduction costs onto the rural, noncapitalist (subsistence) sector. In this context, urban economic growth is attributed to the exploitation of the rural hinterland, and rural underdevelopment is presented as the result of an exploitative process. For the exponents of the Dependencia School, the rural–urban relationship in

[15]For a more recent work on dependency theory, see Mahoney and Rodríguez-Franco (2018).

developing countries reflects the asymmetrical relationships between modes of production.[16]

In contrast to the authors of modernization theories, the representatives of the dependency thesis see rural and urban areas not as separate spheres but as sectors bound together by strongly hierarchical chains of dependence. Despite the emphasis on political-economic interrelation, this way of thinking also conveys a "limited" image of rural and urban areas, in that both are conceived as sharply contrasted subspaces within the national territory. Therefore, rural and urban remained categories for the spatialization of a dualism in the dependency-theoretical school, even if that dualism was no longer conceived as an antithetical, but rather as a complementary dualism or dialectical dichotomy.

2.2.2 Rural/Urban Thinking in Development Politics: Dualistic Persistences

The radical criticism concerning modernization theories also led to a revision of the development strategies which had been propagated up until then and which concentrated primarily on promoting the urban industrial centers. However, it was not primarily the arguments of the Dependencia that led to the emergence of a pro-rural perspective in the development policy discussion, but rather the glaringly obvious and increasingly acute impoverishment in rural areas.

"Urban bias" *and* "rural–urban income gap" *are long-standing central buzzwords in the debate on rural and urban relations in developing countries.*

Lipton (1977) first formulated the thesis of the "*urban bias*" in an influential contribution, showing how the trade, tariff, and tax policies as well as the infrastructure and investment programs of many governments in developing countries favored urban areas at the expense of the poor rural population. The diagnosis: an ever-widening income gap between rural and urban areas. "*Rural–urban income gap*" and *urban bias* became central buzzwords in the development policy discussion on urbanization and "rural exodus" and were often used to characterize the structure of

[16]As such, rural spaces represent the last and weakest link in a worldwide hierarchical chain of exploitation, with the cities in developing countries profiting, as the penultimate link, at least from the exploitation of their rural areas (see Potter 1989; Potter and Unwin 1995).

developing countries. Thus, the dichotomous terms "rich" and "poor" also found their geographical equation in the rural/urban dichotomy.

In response to the criticism of their development models, the international development organizations changed course: Priorities were shifted more to the rural areas, which had been neglected for decades; and from the mid-70s onward, international donor organizations pushed ahead with the strategy of "rural development". This pro-rural view subsequently advanced to become the mainstream of development policy ("ruralization of the development debate"). The prevalent supposition was that, compared to the rural population, urban dwellers had a significantly higher standard of living. The city–country gradient was regarded as the main cause of the rapid urbanization rates in developing countries. The general consensus held that the political "urban bias" was the main cause of the calamity in development, since it had led to a highly inefficient allocation of national resources and thus fed into the tendency toward impoverishment of society as a whole.

"Urban bias" remained a leading concept even as the World Bank and the International Monetary Fund (IMF) reoriented their development strategies, abandoning the Basic Needs Strategy in favor of a neoliberal Structural Adjustment Program (SAP). The implementation of the SAP had severe consequences, especially for the urban population. Analyzing the situation in various African states, Jamal and Weeks (1988) convincingly describe a process they call the "urbanization of poverty," a process which continues to influence development theory discussions on the rural–urban relationship in the Global South to this day. The authors pointed to the following trends, which had become evident since the end of the 1970s (Jamal and Weeks 1988: 273f):

- The real wages of urban workers had been falling dramatically.
- There was no social security nor stability of working arrangements in the formal (urban) economic sector.
- Income disparities between formal and informal labor markets had been narrowing or disappearing altogether.
- The rural–urban income gap had vanished and, in some states, even reversed.
- The national distribution of income had deteriorated in most states.
- Despite the other trends, rural–urban migration had not decreased.

From these observations, Jamal and Weeks (1988) deduced that the distribution of income between rural and urban areas had changed significantly since the 1970s. They criticize that the analyses of poverty and labor market research in Africa had always focused on the relationship between peasant farmers and wage earners, and therefore on the contrast between rural and urban areas, without taking into consideration the differences within these groups. According to Jamal and Weeks (1988), the basic assumption of the widening income gap between the countryside and cities had been conveyed without sufficient empirical evidence. They argued that the resulting research had severely neglected the complexity of interactions between rural and urban areas.

Rural and urban poverty are widely understood as independent phenomena.

With the incipient awareness that, in the course of economic crisis and as a result of the SAP, poverty had entered the towns as well, the emphasis of the development debate shifted yet again, becoming "re-urbanized" to some degree. Thus, the "new" phenomenon of urban poverty gained more attention in development studies, though this did not lead to the abandonment of the rural/urban paradigm. Rural and urban poverty are still often seen as two largely independent phenomena. Dualistic spatial classification has resulted in an overemphasis of urban causes and effects of poverty. Wratten (1995) points out that focusing on the question of whether urban poverty is more extreme than rural poverty distracts from the structural determinants that influence the living conditions of the poor in *both* sectors. An overemphasis of the rural/urban dualism in poverty research had eclipsed the fundamental structural causes. As a result, according to Wratten (1995), the solutions proposed have often been limited to symptom relief at the local (urban *or* rural) level.

In light of this critique, it seems necessary to look for an analytical perspective that regards both the urban and the rural—not as independent spheres, but as parts of one complex system.

2.3 Translocal Perspectives

The translocal perspective serves to overcome conventional container-spatial thinking.

Development research based on dichotomous categories such as modern/traditional, origin/arrival, and urban/rural always focuses on the divisive. Such thinking cannot do justice to the complexity and contextuality of empirical realities. In recent years, therefore, migration and development research have increasingly developed approaches aimed at eroding trivial dualistic categories. This includes the conceptual establishment of a translocal perspective, which, above all, attempts to overcome conventional container-spatial thinking (e.g., Steinbrink 2009a, b; Brickell and Datta 2011; Zoomers and van Westen 2011; Verne 2012; Greiner and Sakdapolrak 2013). In addition to the *Here* and *There*, this perspective pays particular attention to the *In-between*; that is, it focuses on what connects.

Translocality is booming!

Translocal perspectives are currently booming in scholarship. The fact that more and more academics have been working with the concept of translocality over the past twenty years is clearly reflected in the rising number of publications. Figure 2.2 illustrates the temporal development of the search results with the truncated search term "*translocal**" in the database Google Scholar.

Although "translocality" seems to be en vogue, the term is by no means clearly defined. This is mainly because its origin and meaning are quite diffuse and cannot be clearly traced back to the work of a single author. Today, the term is an academic catchword used in very different thematic and disciplinary contexts, for example, in geography (e.g., Lohnert and Steinbrink 2005; Conradson and McKay 2007; Brickell and Datta 2011; Steinbrink 2009a, b; Featherstone 2011; Verne 2012; Gilles 2015), in history and regional studies (e.g., Oakes and Schein 2006; Freitag and von Oppen 2010a), in ethnology and social anthropology (e.g., Appadurai 2003; Argenti and Röschenthaler 2006; Gottowik

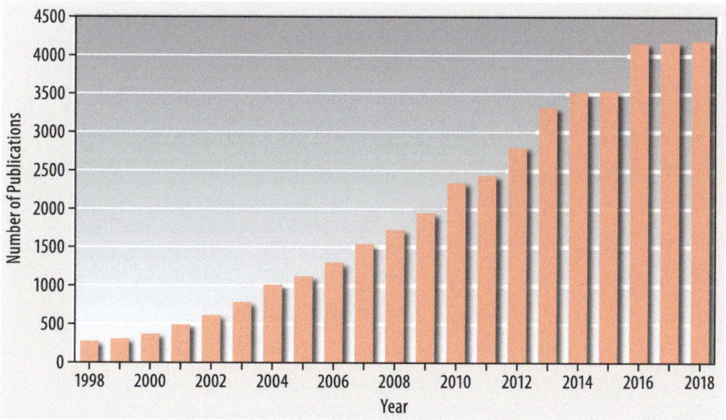

Fig. 2.2 Number of publications on "translocality" per year (1998–2018). *Source* Google Scholar [February 2019], own illustration

2010; Greiner 2010; Marion 2005; Núñez-Madrazo 2007; Peleikis 2003), as well as in development studies (e.g., Zoomers and van Westen 2011).

In the current rapidly growing literature on translocality, one is confronted with an astonishing multitude of other ("big") concepts that cavort in the semantic courtyard of translocality (e.g., mobility, migration, network, social remittances, place, locality, identity, circularity, transfer, [knowledge] flows, social capital). Some authors seek conceptual coherence by relating some of these concepts to existing social theories (e.g., to those of Giddens or Bourdieu: Brickell and Datta 2011; Greiner and Sakdapolrak 2012; Steinbrink 2009a). Some authors, however, use "translocality" more as an "*umbrella term*" (Greiner and Sakdapolrak 2013: 3) to describe different forms of mobility and spatial linkages (e.g., Grillo and Riccio 2004; Ma 2002). Accordingly, Tenhunen (2011: 416) defines "translocality" simply as "*relations that extend beyond the village community*". Freitag and von Oppen (2010b: 5) describe translocality as a "*descriptive tool*" that refers to all phenomena resulting from a series of circularities and forms of transfer between localities. Greiner (2011: 610) explains the term in more detail:

"[...] translocality as the emergence of multidirectional and overlapping networks created by migration that facilitate the circulation of resources, practices and ideas and, thereby, transform the particular localities they connect." (Greiner 2010: 610)

> I define translocality as the emergence of multidirectional and overlapping networks created by migration that facilitate the circulation of resources, practices, and ideas and thereby transform the particular localities they connect. Translocality thus refers to the dynamics, linkages, and interdependencies of the multidimensional social space connecting migrants' areas of origin and destination.

Despite the existing definitional ambiguities and clear differences in the details, some common ground can undoubtedly be found: Across disciplines, authors generally use the concept of translocality when they want to overcome the limitations of conventional concepts based implicitly on specific spatial demarcations and when they wish to consider social phenomena that emerge from the intensive, border- and distance-traversing mobilities of people, goods, information, ideas, and symbols. Anyone who examines the state of literature on translocality will conclude that translocality research primarily seeks to better understand the tensions between mobility and locality arising from increasingly complex socio-spatial dynamics and to consider the realities of the "network society" (Castells 2001) in the "age of migration" (Castles et al. 2013).

Translocality research developed out of discussions around the concept of transnationalism. While it builds on central insights of transnationalism research, the concept of translocality simultaneously aims at overcoming its nation-state focus. For unlike transnationality, translocality does not inherently presume a traversing of national borders. According to Freitag and von Oppen (2010b: 12), translocality instead includes

> a multitude of possible boundaries which might be transgressed, including not limiting itself to political ones, thus recognizing the inability even of modern states to assume, regulate or control movement, and accounting for the agency of a multitude of different actors,[17]

and often, it addresses precisely the special significance of concrete places in the translocal context. Still, the conceptual relationship between translocality and transnationality is ultimately ambiguous: Some authors use the concepts largely synonymously; others try to concretize the concept of transnationality with the help of translocality; and yet others use translocality more as an alternative concept to relativize the importance of the "national".

The focus of translocality research is on spatial mobility and the interconnections between localities.

A common feature of translocality research is that, focusing on the microlevel and "*local-to-local dynamics*," it is almost always based on a relational concept of space. Translocality can be viewed as a research object or a research perspective (Freitag and von Oppen 2010a); but in either case, it is characterized by relationality. The focus is always on the relationships between localities. In other words, translocality encompasses not only the *mobilities* between specific localities but also, and especially, the *interconnections* resulting from various mobilities which form a border- and distance-spanning structure.

Translocality research is a young branch of research that is still in the making. In our opinion, the idea of translocality is a very promising approach, which holds great potential for several reasons. The strengths of the translocality approach are summarized in Fig. 2.3.[18]

[17]See also Gilles 2015: 44–45.

[18]Likewise, see Greiner and Sakdapolrak (2013).

1. **The translocality approach questions and reflects upon dichotomous spatial-geographical concepts** (Agnew 2005) – such as origin and destination, urban and rural, center and periphery – in light of empirical realities (see Steinbrink 2009, Zoomers, van Westen 2011).
2. The actor- and action-oriented focus on the social production and reproduction of **translocality offers insight into the temporal dynamics, path dependencies and spatial-temporal changes of socio-spatial processes.** The concept of translocality is therefore particularly well suited to examining how these processes are socially differentiated and how they affect each other. Consequently, the role of power imbalances in the structuring of "flows and movements" (Massey 1991: 25f) can also be explicitly considered.
3. Translocality opens up an alternative spatial-relational perspective, beyond the usual hierarchy of scale (global – national – regional – local); it emphasizes the special significance of "interlinked places" and "places of networking". Because, unlike the transnationalism approach, the translocality approach relativizes the methodological fixation on the nation state, focusing instead on "local-to-local" dynamics, this perspective may also offer **a new perspective on the "historiography of globalization"** (see Freitag, von Oppen 2010a).
4. Since translocality research explicitly encompasses not only the mobility of people but also flows of money and goods, as well as flows and cycles of ideas, symbols, values, knowledge etc., this approach also provides a stimulating perspective on sedentary sections of the population. As a result, **the structures of interconnetedness regulating the various flows can be analyzed in terms of a coproduction** – a structuring process in which mobile and non-mobile actors, migrants and non-migrants (movers and stayers) respectively, participate.
5. Translocality research is always also locality research, since the respective local conditions and their changes are directly in focus. This is especially true of changes in the physical and natural environment. Translocality can be the result or cause of environmental changes. In this respect, **translocality research can contribute to the discussion on (global) environmental changes** by drawing attention to the central importance of translocal mobility (of people, resources, ideas) and linkages (see Greiner, Peth and Sakdapolrak 2015).

Fig. 2.3 The potential of the translocal perspective. *Source* Own illustration

References

Agnew J (1994) The territorial trap. The geographical assumptions of international relations today. Rev Int Polit Econ 1(1):53–80

Appadurai A (2003) Modernity at large. Cultural dimensions of globalization. University of Minnesota, Minneapolis/London

Argenti N, Röschenthaler U (2006) Introduction between Cameroon and Cuba. Youth, slave trades and translocal memoryscapes. Soc Anthropol 14(1):33–47

Berking H (2006) Raumtheoretische Paradoxien im Globalisierungsdiskurs. In: Berking H (ed) Die Macht des Lokalen in einer Welt ohne Grenzen. Campus, Frankfurt, New York, pp 7–22

Bohle HG (1993) The geography of vulnerable food-systems. In: Bohle HG, Downing TE, Watts MJ (eds) Coping with vulnerability and criticality. Saarbrücken, pp 15–31

Bohle HG (2007) Geographische Entwicklungsforschung. In: Gebhardt H, Glaser R, Radtke U, Reuber P (eds) Geographie. Elsevier, Heidelberg, pp 797–815

Bohle HG (2009) Sustainable livelihood security: evolution and application. In: Brauch HG, Behera N, Kameri-Mbote P et al (eds) Facing global environmental change: environmental, human, energy, food, health and water security concepts. Hexagon series on human and environmental security and peace, vol 4. Springer, Berlin, pp 521–528

Bohle HG, Watts MJ (1993) Hunger famine and the space of vulnerability. GeoJournal 30(2):117–125

Brickell K, Datta A (eds) (2011) Translocal geographies: spaces, places, and connections. Ashgate, Farnham

Castells M (2001) Der Aufstieg der Netzwerkgesellschaft. Das Informationszeitalter 1. Leske und Budrich, Wiesbaden

Castles S, De Haas H, Miller MJ (2013) The age of migration: international population movements in the modern world. Macmillan International Higher Education, London

Chambers R (1989) Vulnerability, coping and policy. IDS Bull 20(2):1–7

Conradson D, Mckay D (2007) Translocal subjectivities: mobility, connection, Emotion. Mobilities 2(2):167–174

Córdova A (1973) Strukturelle Heterogenität und wirtschaftliches Wachstum. Suhrkamp, Frankfurt am Main

De Haan A (2000) Migrants, livelihoods, and rights: the relevance of migration in development policies. Social development working paper 4, London

De Haas H (2010) Migration and development: a theoretical perspective. Int Migrat Rev 1(44):227–264

Dörfler T, Graefe O, Müller-Mahn D (2003) Habitus und Feld. Anregungen für eine Neuorientierung der Geographischen Entwicklungsforschung auf der Grundlage von Bourdieus Theorie der Praxis. Geogr Helv 58:10–23

Etzold B (2017) Mobility, space and livelihood trajectories: new perspectives on migration, translocality and place-making for livelihood studies. In: De Haan L (ed) Livelihoods and development: new perspectives. Brill, Leiden, pp 44–68

Evers T, Wogau P (1973) Dependencia: Lateinamerikanische Beiträge zur Theorie der Unterentwicklung. Das Argument 79:404–454

Faist T (2010) The crucial meso level. Selected studies in international migration and immigrant incorporation, vol 1, pp 59–90

Faist T (ed) (2000) Transstaatliche Räume. Politik, Wirtschaft und Kultur in und zwischen Deutschland und der Türkei. Transcript, Bielefeld

Featherstone D (2011) On assemblage and articulation. Area 43(2):139–142

Frank AG (1969) Die Entwicklung der Unterentwicklung. In: Gunder AG, Echeverria B (eds) Kritik des bürgerlichen Anti-Imperialismus. Rotbuch/Wagenbach, Berlin, pp 30–45

Freitag U, von Oppen A (2010a) Introduction: 'translocality': an approach to connection and transfer in area studies. In: Freitag U, von Oppen A (eds) Translocality. The study of globalising processes from a southern perspective. Brill, Leiden, pp 1–24

Freitag U, von Oppen A (eds) (2010b) Translocality. The study of globalising processes from a southern perspective. Brill, Leiden

Geiger M (2011) Europäische Migrationspolitik und Raumproduktion. Nomos, Baden-Baden

Geiger M, Steinbrink M (2012) Migration & Entwicklung: merging fields in geography. IMIS-Beiträge 42:7–35
Gilles A (2015) Sozialkapital, Translokalität und Wissen: Händlernetzwerke zwischen Afrika und China. Steiner, Stuttgart
Gilman N (2018) Modernization theory never dies. In: HOPE conference: the political economy of development economics: a historical perspective. Duke University Press, Durham, pp 133–151
Glick Schiller N, Basch L, Blanc CS (1997) From immigrant to transmigrant: theorizing transnational migration. In: Pries L (ed) Transnationale Migration. Soziale Welt. Sonderband 12, pp 121–140
Gottowik V (2010) Transnational, translocal, transcultural: some remarks on the relations between Hindu-Balinese and ethnic Chinese in Bali. Sojourn: J Soc Issues Southeast Asia 25(2):178–212
Greiner C (2010) Patterns of translocality: migration, livelihoods and identities in Northwest Namibia. Sociologus 2(60):131–161
Greiner C (2011) Migration, translocal networks and socio-economic stratification in Namibia. Africa 81(4):606–627
Greiner C, Peth S, Sakdapolrak P (2014) Deciphering migration in the age of climate change: towards an understanding of translocal relations in social-ecological systems. TransRe working paper 2, Department of Geography, University of Bonn, Bonn
Greiner C, Sakdapolrak P (2012) Rural–urban migration, agrarian change, and the environment in Kenya: a critical review of the literature. Popul Environ 34(4):524–553
Greiner C, Sakdapolrak P (2013) Translocality: concepts, applications and emerging research perspectives. Geogr Compass 7(5):373–384
Grillo R, Riccio B (2004) Translocal development: Italy–Senegal. Popul Space Place 10(2):99–111
Hein W (1998) Unterentwicklung—Krise der Peripherie. Springer, Opladen
Jamal V, Weeks J (1988) The vanishing rural–urban gap in Sub-Saharan Africa. Int Labour Rev 127:271–292
Kritz MM, Lim LL, Zlotnik H (eds) (1992) International migration systems a global approach. Clarendon Press, Oxford
Lewis A (1954) Economic development with unlimited supplies of labour. Manch Sch Econ Soc Stud 22:139–191
Lipton M (1977) Why poor people stay poor. A study on urban bias in world development. Harvard University Press, London
Lohnert B, Steinbrink M (2005) Rural and urban livelihoods: a translocal perspective. S Afr Geogr J 87:95–103
Ma EK (2002) Translocal spatiality. Int J Cult Stud 5(2):31–152
Mabogunje AL (1970) Systems approach to a theory of rural–urban migration. Geogr Rev 1(2):1–18
Mahoney J, Rodríguez-Franco D (2018) Dependency theory. In: Lancaster C, van de Walle N (eds) The Oxford handbook of the politics of development, pp 1–28
Marion JS (2005) 'Where' is 'there'?: Towards a translocal anthropology. Anthropol News 46(5):18–28
Martin P, Taylor E (1996) The anatomy of a migration hump. In: Martin P, Taylor E (eds) Development strategy, employment, and migration: insights from models. Paris, pp 43–62
Massey D (1991) A global sense of place. Marxism today. Reading human geography. Arnold, London

Massey D, Arango J, Graeme H, Kouaouci A, Pellegrino A, Taylor E (1993) Theories of international migration: a review and appraisal. Popul Dev Rev 19:431–466

Massey D, Arango J, Graeme H, Kouaouci A, Pellegrino A, Taylor E (1994) An evaluation of international migration theory: the North American case. Popul Dev Rev 4(20):699–751

McDowell C, De Haan A (1998) Migration and sustainable livelihoods. IDS working paper 65, Brighton

McGee T (1982) Labour mobility in fragmented labour markets, the role of circulatory migration in rural–urban relations in Asia. In: Safa HI (ed) Towards a political economy of urbanization in third world countries. Oxford University Press, Oxford, pp 57–83

Menzel U (1993) Das Ende der Dritten Welt und das Scheitern der großen Theorie. Suhrkamp, Frankfurt am Main

Müller-Mahn D (1999) Migrationskorridore und transnationale soziale Räume. Eine empirische Skizze zur Süd-Nord-Migration am Beispiel ägyptischer Sans-Papiers in Paris. Abh Anthropogeogr 60:167–200

Núñez-Madrazo C (2007) Living 'here and there': new migration of translocal workers from Veracruz to the Southeastern United states. Anthropol Work Rev 28(3):1–6

Oakes T, Schein L (2006) Preface. In: Oakes T, Schein L (eds) Translocal China: linkages, identities and the reimaging of space. Routledge, London, pp 12–13

Palloni A, Massey DS, Ceballos M, Espinosa K, Spittel M (2001) Social capital and international migration: a test using information on family networks. Am J Sociol 106(5):1262–1298

Peleikis A (2003) Lebanese in motion: gender and the making of a translocal village. Transcript, Bielefeld

Pieterse JN (2010) Development theory: deconstructions/ reconstructions. SAGE Publications Ltd, London

Potter R (1989) Urban–rural interaction, spatial polarisation and development planning. In: Potter R, Unwin T (eds) The geography of urban–rural interaction in developing countries. Routledge, London, pp 323–333

Potter R, Unwin T (1995) Urban rural interaction: physical form and political process in the third world. Cities 12:67–74

Pries L (2008) Die Transnationalisierung der sozialen Welt. Suhrkamp, Frankfurt

Scholz F (2004) Geographische Entwicklungsforschung. Teubner, Berlin

Scoones I (2009) Livelihoods perspectives and rural development. J Peasant Stud 36(1):171–196

Senghaas D (1972) Imperialismus und strukturelle Gewalt. Analysen über abhängige Reproduktion. Suhrkamp, Frankfurt am Main

Senghaas D (1974) Peripherer Kapitalismus. Analysen über Abhängigkeit und Unterentwicklung. Suhrkamp, Frankfurt am Main

Standing G (1985) Labour circulation and the labour process. Croom Helm, London

Stark O (1991) The migration of labour. Blackwell, Cambridge

Stark O, Bloom D (1985) The new economics of labour migration. Am Econ Rev 75:173–178

Steinbrink M (2009a) Leben zwischen Land und Stadt, Migration, Translokalität und Verwundbarkeit in Südafrika. Springer, Wiesbaden

Steinbrink M (2009b) Urbanisation, poverty and translocality insights from South Africa. Afr Popul Stud 23:219–252

Steinbrink M (2012) Migration, Netzwerk und 'Entwicklung'. Translokale Livelihoods in Südafrika. IMIS-Beiträge 42:165–204

Steinbrink M, Peth S (2014) Hier, dort und dazwischen. Geogr Rundsch 74(11):32–39

Steinbrink M, Schmiz A (2019) Migration & Entwicklung—neue Debatten um ein altes Thema. In: Gamerith W, Scharfenort N (eds) Menschen, Migration und Mobilität. Selbstverlag Fach Geographie der Universität Passau, Passau (forthcoming)

Tenhunen S (2011) Culture, conflict, and translocal communication: mobile technology and politics in rural West Bengal, India. Ethnos 76(3):398–420

Thieme S (2008) Sustaining livelihoods in multilocal settings: possible theoretical linkages between livelihoods and transnational migration research. Mobilities 3:51–71

Todaro M (1969) A model of labour migration and urban unemployment in less developed countries. Am Econ Rev 59:138–148

Verne J (2012) Living translocality: space, culture and economy in contemporary swahili trading connections. Steiner, Stuttgart

Waddington H, Sabates-Wheeler R (2003) How does poverty affect migration choice? A review of literature. IDS working paper T3, Brighton

Wenzel HJ (2012) Migration in der Geographischen Entwicklungsforschung: Konturen und Verknüpfungen unter Bezugnahme auf Studien zu Afrika. IMIS-Beiträge 42:37–60

Wimmer A, Glick Schiller N (2002) Methodological nationalism and beyond: nation-state building, migration and the social sciences. Glob Netw 2(4):301–334

Wolpe H (1972) Capitalism and cheap labour power in South Africa. Econ Soc 1:425–456

Wratten E (1995) Conceptualizing urban poverty. Environ Urban 7 (1):11–37

Zoomers A, van Westen G (2011) Introduction: translocal development, development corridors and development chains. Int Dev Plan Rev 33 (4):377–388

Chapter 3
Translocal Livelihoods: New Perspectives in Livelihood Research

Abstract The most advanced operationalization of "translocality" is found in (geographical) development studies, specifically within livelihood research. The livelihood perspective and the translocal perspective are analytically mutually enriching. On the one hand, the livelihood approach opens a methodological perspective on translocal agency, allowing for the investigation of mobility and space-spanning structures. On the other hand, the translocal perspective allows us to better understand the role of migration and space-spanning social networks for livelihood security. In this chapter, we first explain why the *translocalization of the livelihood perspective* is analytically necessary (Sect. 3.1), before discussing some important concepts and *methodological implications* of the translocal livelihood approach (Sect. 3.2). We introduce the terminology of "translocal social space" (Sect. 3.2.1), "translocal community" (Sect. 3.2.2), and "translocal household" (Sect. 3.2.3) and examine the necessity of a "bi- or multilocal approach to fieldwork" (Sect. 3.2.4). These concepts are useful for overcoming the local bias of conventional concepts within livelihood research.

Keyword Translocal livelihoods · Translocal community · Translocal household · Translocal methodology

The discussion of translocality is particularly well advanced within (geographical) development studies, with the most mature operationalization of the concept found in the context of livelihood research. Various authors have shown that the livelihood perspective and the translocal perspective are analytically mutually enriching. On the one hand, the analysis of

M. Steinbrink and H. Niedenführ, *Africa on the Move*, Springer Geography,
https://doi.org/10.1007/978-3-030-22841-5_3

livelihood systems (or livelihood strategies) opens an analytical and methodological perspective on translocal action, allowing for the investigation of the space-spanning structures. On the other hand, the translocal perspective allows us to more precisely understand the great importance of migration and space-spanning social networks for livelihood security.

In the following, we first explain why the translocalization of the livelihood perspective is analytically advantageous (Sect. 3.1), before introducing the theoretical concept of translocal livelihoods and discussing some methodological implications of this approach (Sect. 3.2). Specifically, we introduce the terminology of "translocal social space" (Sect. 3.2.1), "translocal community" (Sect. 3.2.2) and the "translocal household" (Sect. 3.2.3) and examine the necessity of a "bi- or multilocal approach to fieldwork" (Sect. 3.2.4).

These concepts are meant to help researchers overcome the local bias of conventional concepts within livelihood research.

3.1 Translocalizing the Livelihood Perspective

Livelihood research focuses on the micro-level of households: their available resources, capabilities, options and strategies in specific contexts of action.

For almost three decades, livelihood approaches (Carney 1998; Chambers 1987; Chambers and Conway 1992; DFID 1999; Scoones 1998) have been a popular analytical framework for investigating the strategic action of people in rural and urban contexts marked by risk and existential uncertainty. As an action- and actor-oriented analytical instrument, livelihood research focuses on the micro-level of households, including the diverse material and immaterial resources (*livelihood assets*) available to them (Geiser 2017: 13), as well as their options and strategies for using these resources (*livelihood strategies*) in specific contexts of action (*vulnerability context, transforming structures and processes*). The strength of the approach is that it enables a deeper understanding of the rationalities behind the decision-making and concrete strategic actions of household members. It analyzes how actors strategically use household resources to adapt to or cope with the challenges of everyday life. By concentrating on the daily life practices of potential target groups, the livelihood approach's analysis lends itself to the work of development cooperation practitioners and provides clear starting points for external interventions. It may be because of this practical applicability—particularly in the context of development projects—that empirical livelihood studies have all too often remained very descriptive. In this respect, the livelihood approach is ultimately less about explaining the emergence of existentially threatening situations than about describing the situation (*vulnerability context*) and how people (can) react to it according to their available bundle of resources.

Many development researchers regard the livelihood approach as an indispensable tool for the analysis of rural (e.g. De Haan 2012; De Haas 2010; King 2011; Mazibuko 2012; Scoones 2009) and urban (Krüger 1997, Farrington et al. 2002; Lohnert and Steinbrink 2005; Rakodi 1995, 2002; Dasgupta and Lloyd-Jones 2018) dynamics at the micro level. "As a matter of fact, livelihood research has become an integral part of contemporary attempts to understand why some people are in a position to maintain or even improve their living standards while others not" (Geiser 2017: 14). This estimation of the approach's potential remains high despite the criticism it has received in the last two decades regarding its theoretical and heuristic limitations. One major point of criticism refers to the approach's conceptual weakness when it comes to considering macrostructural factors (e.g. Dörfler et al. 2003; Steinbrink 2009a, b; Müller-Mahn and Verne 2010; Scoones 2009) and social power relations beyond the household level (e.g. De Haan 2008, 2012). Critics often point out that this under-complexity makes it necessary to link the livelihood-concept with other theoretical approaches (Bebbington and Batterbury 2001; De Haan 2012; Thieme 2008; van Dijk 2011; Greiner and Sakdapolrak 2013).

Another conceptual weakness of conventional livelihood research relates to the "container-spatial bias" (Steinbrink 2009a, b) of the research perspective.

> 'Classical' livelihood approaches largely failed to grasp spatial mobility and migration dynamics in rural households, family systems and communities. (Benz 2014: 260)

Conventional livelihood research has long been marked by "container-spatial bias."

Although livelihood research has long and intensively dealt with *sectoral income diversification*, the approach has hardly been able to capture the *spatial diversification of livelihood systems* (King 2011; Steinbrink 2012; Etzold 2017). Since the empirical research has predominantly been based on (single-site) local or regional case studies, investigators have often failed to understand that the observed circumstances of local life are merely partial structures of the overarching context, nor have they adequately considered that the investigated groups frequently only constitute substructures of communities that organize themselves translocally. Such livelihood research, which is oriented toward territorial boundaries, too often fragments the social contexts in question, especially as migration and social interconnectedness across vast distances increasingly form the basis of people´s livelihood security. In order to analyze livelihood systems and the vulnerability of those groups dependent upon them, the investigation of migratory movements and social networks must become a central element of research work.

Conventional livelihood studies have had great difficulty in conceptualizing the new mobilities in the rural periphery of the Global South. Only in the last ten years have efforts been made in international development research to take greater account of the spatial (or space-spanning) dimension of livelihoods (King 2011). The perspective of livelihood research is thus ripe for a shift to free itself from the container-spatial thinking of previous research. When it comes to fleshing out an alternative or expanded conceptual approach, the focus must lie first and foremost on developing a *translocal perspective in livelihood research.*

Many names, (more or less) one meaning: "multilocal livelihoods," "multilocational livelihoods," "translocal livelihoods," "transnational livelihoods".

In this context, some authors use the term "*multilocal livelihoods*" (De Haan 2008; De Haan and Zoomers 2003; Elmhirst 2012; Thieme 2008) or "*multilocational livelihoods*" (Deshingkar and Farrington 2009; Ramisch 2014, 2015), while others speak of "*translocal livelihoods*" (Lohnert and Steinbrink 2005; Steinbrink 2007, 2012; Long 2008; Greiner 2010, 2012) or—if a national border is crossed—of "*transnational livelihoods*" (Bebbington and Batterbury 2001). What unites these concepts is that they are based on a relational perspective, i.e. an understanding of networks and interactions that span spatial distances. Developing the translocal livelihood perspective still requires some modifications to the conceptual instruments and the empirical research design (see Lohnert and Steinbrink 2005: 97ff), as we will discuss in the next chapter.

3.2 Methodological Implications

The challenge in combining the translocal perspective and livelihood research lies in the tendency of the livelihood approach to get stuck in the territorial trap discussed in Chap. 2. Through a kind of spatializing automatism, livelihood studies' strong focus on action and thus on the micro-level of the actors, households and communities has all too easily led researchers to equate the micro-level with the local level. They have often presumed the livelihood systems they analyze to be locally bounded, that is limited to one locality, and have consequently focused their examinations solely on "the local."

Consequently, the translocalization of the livelihood perspective has some far-reaching implications.[1] On the one hand, there is a need to reconceptualize the central units of analysis "*community*" (see Sect. 3.2.2) and "*household*" (see Sect. 3.2.3), and on the other hand, there is a need to adapt the methodology of field studies (*bi- or multilocal approach*; see Sect. 3.2.4). But before discussing these terms and approaches, we must first grasp the concept of *translocal social space* (see Sect. 3.2.1).

3.2.1 *Translocal Social Space*

The idea of *social space* is central to conceptualizing translocality. For our purposes, we understand social space as a compact nexus of social interlinkages between actors. Initially, social space is "non-geographical," it does not depend on spatial demarcations; only when we take the concept a step further and address the whereabouts of actors does it gain "geographical relevance": If socially interlinked actors find themselves at different localities, we can speak of *translocal social space*. Following Ludger Pries' (2001) definition of *transnational social space*, *translocal social space* can be understood as pluri-local, interpersonal links which constitute a relatively stable, compact configuration of daily life practices. *Translocal social space* spans over several localities and transcend spatial demarcations. That makes

[1]For more on the concepts "translocal household" and "translocal community" presented here, as well as the necessity of bi- or multilocal field research, see Steinbrink (2009a, b: 45–52).

analyzing the translocal practices of everyday life indispensable to the examination of socio-spatial linkages.

In many African countries, daily life practices can all too often be equated with practices aimed at daily survival. Thus, to understand *translocal social space* in Africa and analyze the socio-spatial linkages that constitute it, it makes sense to turn to livelihood and vulnerability studies, which focuses explicitly on strategic action within the context of existential risk, crisis and uncertainty.

3.2.2 Translocal Community

Translocal Community: *A community whose members live in different places and are linked to one another by functional dependencies that tend to be stronger than those that connect them to other people in the wider social environment. The community is thus to be understood as a social network and not as a place.*

In order to develop a translocal perspective, livelihood and vulnerability research must somehow break with the tradition of social-anthropological *community studies*, which is fundamentally based on a notion of congruence between social space and territory.

Within classic community studies, the concept of community has generally been conceived as a locally bounded entity which is often assumed to have a clearly definable territory and comprise a homogenous and relatively closed social formation, characterized by personal relationships, mutual responsibility and solidarity. Empirical research into social vulnerability based upon local community studies continuously reproduce the image of the community as a local interweaving of culture and identity—a geographically and socially bound group of people characterized by immobility, limited radii of action, intensive intragroup communication and a dense structure which produces conformity (Lohnert and Steinbrink 2005: 97). These studies often remain conceptually trapped in geographical space and tend to assert locality as a precondition for sociality. Livelihood security is often seen as socially embedded in a moral economy, within which the survival of the poor depends upon the norms of solidarity and reciprocity. However, conventional approaches to community imagine these social networks primarily as *local* interweavings of relationships. Reciprocity and mutuality are often taken for granted or considered implicitly as the product of intensive local social interaction within geographically bound social networks.

This local fixation of conventional livelihood studies' analytical concepts makes integrating social interaction beyond the local context exceedingly difficult, leading them to be largely neglected. Because the social entities analyzed ("*households*"; see below) are defined via their locality, livelihood systems are mostly interpreted as locally bound. In many cases, migration is regarded as an anomaly (in cases of crisis or disaster) and not as an integral part of livelihood itself.

In view of today's migration processes and the great importance of social networks that transcend community boundaries, these presuppositions and the container-spatial thinking of classical community studies are clearly losing their empirical plausibility.

To develop a translocal perspective, we need to develop the notion of "*translocal community*." Translocalizing the community concept allows us to account for the fact that the social context has been decoupled from the actors' place(s) of residence. Despite the limitations of the term as applied by conventional approaches, "community"—with its spatial connotations—is still appropriate, because the translocal community is neither completely "de-spatialized" nor "de-localized" but does indeed show geographical references. However, instead of being locally bound, the community is translocally organized. Expanding on Elias' (1974) general definition of community, a *translocal community* is (i) a group of (translocal) households (see below), whose members (ii) live in diverse locations, which are (iii) connected through functional interdependencies, which tend to be (iv) stronger than those interdependencies which connect them to other people in their wider social environment, and who (v) use a common origin as a symbolic link and the basis for a collective identity.[2]

The translocal community should be regarded as an (intermediate) result of a continuous process in the collective construction of identity around the idea of a common origin. In this process, "home" is socially constructed as a quasi-spatial anchor for norms, values and worldviews, which are reproduced *by* the community members as the norms, values, worldviews *of* the community. Accordingly, we can understand the community not as a location but as a social network primarily defined by what people do for each other and not where they live. The locality of the home village ("home") is socially constructed within the network context as a criterion of inclusion and exclusion ("in- and out-group") and is used to set social boundaries. Through the

---

[2]For an instructive critique of the community concept, see Brint (2001).

medium of common origin, local and translocal relationships are established and maintained.

At the same time, the different places where community members reside represent the arenas in which (translocal) action is taken and resources are used. Each site offers its own set of opportunity structures, which the translocal network links together, merging them into a single opportunity structure of the translocal social space. Moreover, the places of residence are also sites of face-to-face contact, where direct interaction, social control and sanctioning can take place and where community links can be continuously reproduced in communication and interaction.

Ultimately, the concept of the *translocal community* is largely congruent with the more abstract concept of *translocal (social)* space (see Sect. 3.2.1), but it refers more strongly to the special significance of communality as a resource.

In order to understand the problems of poverty and vulnerability in Africa—"home to most of the deeply poor," according tothe World Bank and IMF (De Haan 2017: 4)—we need to examine not only the relevance of translocal communality for securing livelihoods but also the extent to which the translocality of communities influences livelihoods. Answering the questions of why and how these translocal communities emerge, to which social dynamics they are subject and how they are maintained, changed or dissolved can contribute significantly to our understanding of rural-urban relations and migration and what they mean for the livelihood security in (rural *and* urban) Africa.

3.2.3 The Translocal Household

The second challenge in developing a translocal perspective in livelihood research is the reconceptualization of the term household: the household is *the* central unit of observation and analysis for livelihood research. As a social unit, the household is generally regarded as the carrier of the livelihood system.[3]

[3]However, there is also well-founded criticism of the household approach, which "may mask intra-household power inequalities (Rodenburg 1997) and largely ignores individual decision-making and the influence of non-household members" (De Haas and van Rooij 2010: 45).

Most of the numerous household definitions found in the literature agree on the following central points: a household is defined as a group of people who make coordinated decisions about the use of resources, pool their income and *live together* (Ellis 1998; Barrett et al. 2001).[4] Here again, we see the container-spatial bias of conventional perspectives manifested in the defining criterion of coresidence. Underlying this fundamental criterion is the implicit assumption that the relationships of exchange and cooperation among the occupants of a house are automatically stronger and more important than their relationships to people who live elsewhere.

Necessary reconceptualization of the household concept: liberation from the "four walls".

A translocal perspective, by contrast, requires us to reconceptualize the category of the household as a *consumption and reproduction collective* (Meillassoux 1983), liberating it from its "four walls." The household can be then defined not by its members residing in a shared abode but by their reciprocal relationships of sharing, cooperation and exchange, which are governed by specific social roles and cultural norms. In other words, the household—like the translocal community (see above)—must be defined by what people do for each other and not by where they eat or sleep.

It therefore makes sense to define a household as a socially recognized group, whose members coordinate their activities of consumption, reproduction and resource use over a long period. In consequence, the household members need not

[4]Quite a few authors point out that, in practice, it is extremely difficult to clearly identify the limits of a household according to the defining criteria above and to distinguish it from other organizational units. They underline that consumption within residential groups is not always collective, nor are all members' incomes necessarily contributed to a joint budget. And decisions about the use of resources are often primarily made by the heads of household alone. Residential groups are not static but rather highly flexible structures, whose composition constantly changes. Nor are these groups to be regarded as isolated, as they are interlocked in many ways.

The structures and compositions of the residential group defined as a household vary greatly across time and space, according to social, economic and cultural conditions (see Russell 1993). The idea of the household as a universally comparable analysis category is thus highly questionable.

necessarily live together![5] If the members of a thus defined household live in different places and organize their living jointly over a spatial distance, we can speak of a *translocal household*.[6] This understanding of the term accounts for the fact that the organization of livelihood is often not spatially bound, but that it is the translocality of the household unit, which forms a basis for the existence of its members. In fact, livelihood security is often only possible through a way of life that implies multiple localities.

Translocal Household: *A socially recognized, jointly economizing collective, whose members do not permanently live in one place, but do coordinate their activities of consumption, reproduction and resource use over a long period of time.*

Within translocal households, numerous social, emotional and cultural ties exist among actors at different locations, as do diverse exchange and cooperation relationships that take the form of transfers of money, goods, people and information. For an exemplary schematic compilation see Table 3.1.

Viewed from this perspective, translocal task-sharing is more a division of functions between different household locations than a strict distribution of tasks among household members, who often—for longer or shorter periods—live either in the city or in the countryside, where they take on site-specific tasks respectively. Thus, in the context of translocal households, rural and urban household locations have characteristic functions. Figure 3.1 illustrates the translocal division of functions.

---

[5]Thus, the house itself is no longer the defining criterion; it is instead, first, an economic resource of the household collective and, second, the point of reference for constructing a common identity. Membership is linked to the recognition of social norms and to certain rights and obligations within the group. At the same time, membership in this household collective can regulate the incorporation within larger social contexts (e.g. clan or community).

[6]The fundamental bases of the concept "translocal household" (Steinbrink 2009a, b) can also be found in other authors' household definitions. Vorlaufer (1992), Frayne and Pendelton (2001) and Muzidziwa (2001) speak of "gespaltenen Haushalten"/"split households," Tacoli (1998) of "multi-spatial households" or "multi-locational" and "multi-sectoral households," Forsberg (1999) speaks of "satellite households" and Lohnert (2002) of "multi-local households," Adepoju (1995) of "dual households," Smith (1998) of "multiple home households." Cowen and Kinyanjui (1977), Kitching (1980) and Orvis (1997) again use the term "straddling households" (see also Ramisch 2015). All these terms refer to the phenomenon of spatially fragmented household units. Here, though, we will refrain from discussing the similarities and differences of the respective concepts. On the household concept, see also Schmidt-Kallert (2009); for a critique of these concepts, see Greiner (2012).

Table 3.1 Types and direction of exchange within translocal households in an urban-rural context

Urban		Rural
Money		
– financial support for schoolchildren – "starter money" for newly arrived migrants	→ ←	**Regular monetary transfers (remittances)**: – for consumer purposes – for investment purposes (increasingly via mobile phone [mobile cash]) **Special transfers**: – in acute emergency situations (e.g. illness) – school fees and uniforms – financing for celebrations and ceremonies (e.g. weddings, circumcisions, funerals) – bride price ("Lobola") – investments in house construction – investment in livestock
Goods		
– food products – traditional medicines	→ ←	– Electronic devices (televisions, mobile phones, radio, etc.) – furniture – clothing, shoes – food products – small gifts (alcohol, sweets, etc.)
People		
Rural-urban migration: – labor migration – education migration – health migration – partner migration (reunification) – co-migration of small children – flight from (domestic) violence **Visits**: – during school holidays – "sounding out" livelihood possibilities	↔ ↔	**Return migration (urban-rural)**: – in the event of disability – in old age – in cases of serious illness – in cases of pregnancy/ during early motherhood – in cases of prolonged unemployment – in times of high labor demand in agriculture (e.g. harvest period) – for marriage (largely women only) – return of young children born in the city – in the event of failure at school

(continued)

Table 3.1 (continued)

Urban		Rural
		Visits: – for vacation – during school holidays – for festivals and rituals (funerals, weddings, circumcisions, etc.)
Information		
– about the situation of relatives in the countryside (health, school, income, agriculture, conflicts, problems, hardships, etc.) – about the situation in the village in general (deaths, weddings, celebrations, rituals, problems/conflicts, gossip, new opportunities, special events, weather conditions, etc.)	↔	– about family members in the city (health, work, housing, income, problems, disputes, etc.) – about other migrants from the village in the city – about job opportunities for potential migrants – about living conditions/ possibilities – about access to educational institutions – about access to healthcare facilities – about special events in the city

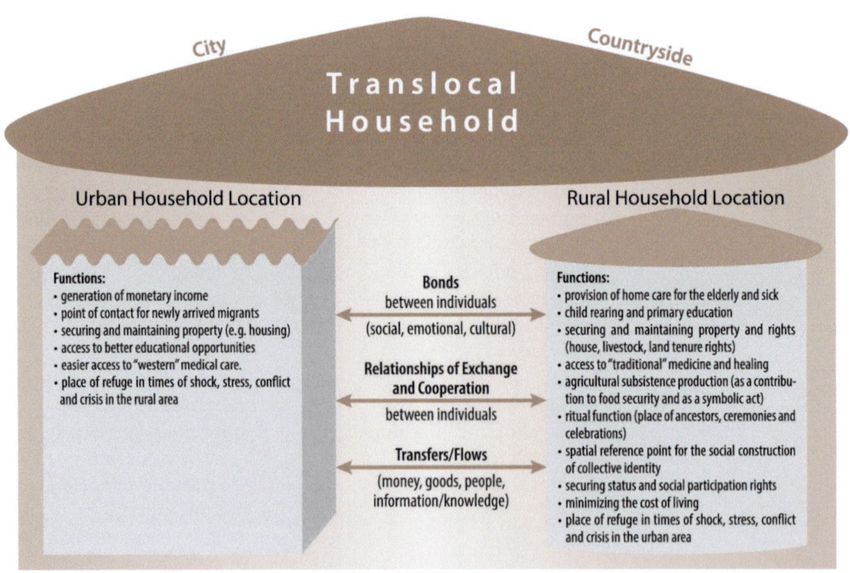

Fig. 3.1 Functions of rural and urban locations of translocal households. *Source* Own illustration

With the concept of the translocal household, the most important term for a translocal perspective of livelihood research is now at our disposal.

Translocal Livelihood System: *A system ensuring the subsistence of a translocal household; it comprises the coordinated use of the tangible and intangible resources available at different household locations as well as the coordination of strategic action with the aim of satisfying (individual and collective) needs and securing long-term subsistence.*

Translocal households are the carriers of *translocal livelihood systems*. As such, translocal livelihood systems are systems for ensuring the subsistence of a household whose members do not all reside in the same place. In a translocal livelihood system, both the use of the available tangible and intangible assets and the strategic actions of the members at the different household locations are coordinated in order to satisfy individual and collective needs and secure the long-term subsistence of all household members.[7]

With the help of this definition, translocal action (e.g. between rural and urban areas in Africa) can be conceptualized as strategic action and an integral component of collective livelihood security (Steinbrink 2009a, b: 90).

The actors at the various household locations are connected through interlinkages that transcend container-spatial boundaries, and they depend on one another. The structure and organization of translocal households illustrate this relationship of existential dependency among the actors at different locations. As such, urban and rural livelihoods are often only elements of a translocal livelihood system: Spatial diversification allows household members to combine economic activities, taking advantage of the opportunities available at different locations. Through translocal networks, the opportunity structures at the different locations merge into one single opportunity structure within the *translocal social space*. Migration, cooperation and exchange are prerequisite conditions for the functioning of this translocal livelihood system.

3.2.4 A Bi- or Multilocal Research Approach

Conventional, stationary field studies are unable to meet the translocal perspective's demands on the methodological

[7]This definition of "translocal livelihood system" was developed based on the definitions of "livelihood system" by Chambers and Conway (1992) and "livelihood" by Grawert (1998) and Carney (1998), respectively.

design of research. Though mobility can be a subject of observation within stationary (single-site) field studies, the methodological focus of such research remains fixed on the concrete place and the people who can be found there. Conventional field studies essentially preassume settledness (sedentarism) as the norm, interpreting migration as the exception to the rule in otherwise sedentary lives. It is precisely this basic assumption that the translocal perspective abandons and even partly reverses. Both spatial mobility and settledness are seen as normal.

This methodological reversal puts the researcher under mobility pressure: "translocality researchers" must always be "mobility researchers," since migration/mobility are preconditions of translocality and integral aspects of the organization of livelihood. Adapting the research design to translocal realities also entails undertaking empirical work in bi- or multilocal contexts.[8] The realization that territorial community boundaries do not necessarily represent the boundaries of the analyzed group's social space requires the researcher at the beginning of his/her field work to inquire which localities are socio-spatially interconnected. The search for the (social) space of analysis becomes part of the research itself. Thus, the researcher is initially an explorer, embarking on a journey into a "*socio-spatial terra incognita.*" It is the studied groups themselves who lead the researcher on the path to the hitherto blank spaces of the research map, as they show him/her "their social space" within the translocalized social context. Translocal livelihood studies center on the analysis of those social relations between people in different places, and thus on the social links between localities that are meaningful for livelihood security. It is therefore indispensable that researchers do their fieldwork at multiple sites.

References

Adepoju A (1995) Migration in Africa. An overview. In: Baker J, Tade A (eds) The migration experience in Africa. Uppsala, Nordiska Afrikainstitutet, pp 87–108

[8]Within the context of cultural anthropology, Marcus (1995) refers to this kind of mobile fieldwork as "multi-sited ethnography."

Barrett C, Reardon T, Webb P (2001) Nonfarm income diversification and household livelihood strategies in rural Africa: concepts dynamics and policy implications. Food Policy 26(4):315–331

Bebbington AJ, Batterbury SPJ (2001) Transnational livelihoods and landscapes: political ecologies of globalization. Ecumene 8(4):369–380

Benz A (2014) Mobility, multilocality and translocal development: changing livelihoods in the Karakoram. Geogr Helv 69:259–270

Brint S (2001) Gemeinschaft revisited: a critique and reconstruction of the community concept. Sociol Theory 19(1):1–23

Carney D (1998) Sustainable rural livelihoods: what contribution can we make?. Department for International Development, London

Chambers R (1987) Sustainable livelihoods, environment and development: putting poor rural people first. Discussion paper 240, IDS Institute of Development Studies, Brighton

Chambers R, Conway G (1992) Sustainable rural livelihoods: practical concepts for the 21st century. Discussion paper 296, IDS Institute of Development Studies, Brighton

Cowen MP, Kinyanjui K (1977) Some problems of capital and class in Kenya. Occasional paper 26 Nairobi, Institute for Development Studies. Available via OpenDocs. http://opendocs.ids.ac.uk/opendocs/handle/123456789/795. Accessed 18 Mar 2019

Dasgupta N, Lloyd-Jones T (2018) Heterogeneity and vulnerability in the urban informal economy: reworking the problem in the current context. The case of Uganda. World Dev Perspect 10:64–72

De Haan A, Zoomers A (2003) Development geography at the crossroads of livelihood and globalization. Tijdschrift voor economische en sociale geografie 94(3):350–362

De Haan L (2008) Livelihoods and globalisation. Paper presented at the international conference on research for development, Bern, 2–4 July 2008

De Haan L (2012) The livelihood approach: a critical exploration. Erdkunde 66:345–357

De Haan L (2017) From poverty to social exclusion: a livelihoods introductory. In: De Haan L (ed) Livelihoods and development: new perspectives. Brill, Leiden, pp 1–12

De Haas H (2010) Migration and development: a theoretical perspective. Int Mig Rev 1(44):227–264

De Haas H, van Rooij A (2010) Migration as emancipation? The impact of internal and international migration on the position of women in rural Morocco. Oxf Dev Stud 38:43–62

Deshingkar P, Farrington J (2009) Circular migration and multilocational livelihood strategies in rural India, New Delhi. University Press, Oxford

DFID (Department for International Development) (1999) Sustainable livelihoods guidance sheets. Available via Department for International Development, London. http://www.livelihoodscentre.org/documents/20720/100145/Sustainable+livelihoods+guidance+sheets/8f35b59f-8207-43fc-8b99-df75d3000e86. Accessed 18 Mar 2019

Dörfler T, Graefe O, Müller-Mahn D (2003) Habitus und Feld. Anregungen für eine Neuorientierung der Geographischen Entwicklungsforschung auf der Grundlage von Bourdieus Theorie der Praxis. Geogr Helv 58:10–23

Elias N (1974) Towards a theory of communities. In: Bell C, Newby H (eds) The sociology of community: a selection of readings. Frank Cass & Co, London, pp 9–41
Ellis F (1998) Survey article household strategies and rural livelihood diversification. J Dev Stud 35(1):1–38
Elmhirst R (2012) Displacement, resettlement, and multi-local livelihoods. Critical Asian Stud 44:131–152
Etzold B (2017) Mobility, space and livelihood trajectories: new perspectives on migration, translocality and place-making for livelihood studies. In: De Haan L (ed) Livelihoods and development: new perspectives. Brill, Leiden, pp 44–68
Farrington J, Ramasut T, Walker J (2002) Sustainable livelihoods approaches in urban areas: general lessons with illustrations from Indian cases. ODI working paper 162, Overseas Development Institute, London
Forsberg J (1999) Provisioning by rural-urban interaction and mobility adjusted livelihoods and social relations in Mvumi, Tanzania. Paper presented at the conference 'Between town & country livelihoods settlement and identity formation in Sub-Saharan Africa', Rhodes University, East London, 27–30 June 1999
Frayne B, Pendleton WC (2001) Migration in Namibia: combining macro and micro approaches to research design and analysis. Int Migr Rev 35(4):1054–1085
Geiser U (2017) Understanding poverty, defining interventions: why social relations need more attention in livelihoods analyses and why this complicates development practice. In: De Haan L (ed) Livelihoods and development: new perspectives. Brill, Leiden, Boston, pp 13–43
Grawert E (1998) Der, 'Livelihood Approach': Eine Analysemethode für komplexe sozioökonomische Absicherungsformen dargestellt am Beispiel Westsudan. Peripherie 69:67–87
Greiner C (2010) Patterns of translocality: migration, livelihoods and identities in Northwest Namibia. Sociologus 2(60):131–161
Greiner C (2012) Can households be multilocal? Conceptual and methodological considerations based on a Namibian case study. Erde 143:195–212
Greiner C, Sakdapolrak P (2013) Translocality: concepts, applications and emerging research perspectives. Geogr Compass 7(5):373–384
King B (2011) Spatialising livelihoods: resource access and livelihood spaces in South Africa. Trans Inst Br Geogr 36:297–313
Kitching G (1980) Class and economic change in Kenya: the making of an African Petit-Bourgeoisie. Heinemann, London
Krüger F (1997) Urbanisierung und Verwundbarkeit in Botswana. Centaurus, Pfaffenweiler
Lohnert B (2002) Vom Hüttendorf zur Eigenheimsiedlung: Selbsthilfe im städtischen Wohnungsbau: Ist Kapstadt das Modell für das neue Südafrika?. Universitätsverlag Rasch, Osnabrück
Lohnert B, Steinbrink M (2005) Rural and urban livelihoods: a translocal perspective. S Afr Geogr J 87:95–103
Long N (2008) Translocal livelihoods, networks of family and community and remittances in Central Peru. In: DeWind J, Holdaway J (eds) Migration and development within and across borders: research and policy perspectives on internal and

international migration. International Organization for Migration, Genf, pp 39–70

Marcus GE (1995) Ethnography in/of the world system: the emergence of multi–sited ethnography. Annu Rev Anthropol 24(1):95–117

Mazibuko S (2012) Understanding underdevelopment through the sustainable livelihoods approach. Community Dev 44:173–187

Meillassoux C (1983) The economic bases of demographic reproduction: from the domestic mode of production to wage-earning. J Peasant Stud 11(1):50–61

Müller-Mahn D, Verne J (2010) Geographische Entwicklungsforschung – alte Probleme neue Perspektive. Geographische Rundschau 62 (10):4–11

Muzidziwa VN (2001) Keeping a foot in the village: Masvingo urban women. J Soc Dev Afr 16(1):85–99

Orvis SW (1997) The Agrarian question in Kenya. University of Florida Press, Gainesville

Pries L (2001) Internationale Migrationen. Soziologische Themen, Transcript, Bielefeld

Rakodi C (1995) The household strategies of the urban poor: coping with poverty and recession in Gweru, Zimbabwe. Habitat Int 19:447–471

Rakodi C (ed) (2002) Urban livelihoods: a people-centred approach to reducing poverty. Earthscan, London

Ramisch J (2014) 'We will not farm like our fathers did' multilocational livelihoods cellphones and the continuing challenge of rural development in Western Kenya. In: Sick D (ed) Rural livelihoods regional economies and processes of change. Routledge, London, pp 10–25

Ramisch J (2015) 'Never at Ease': cellphones multilocational livelihoods and the metabolic rift in Western Kenya. Agric Hum Values 33(4):979–995

Russell M (1993) Are households universal? On Misunderstanding Domestic Groups in Swaziland. Dev Change 24(4):755–785

Schmidt-Kallert E (2009) A new paradigm of urban transition: tracing the livelihood strategies of multi-locational households. Erde 140:319–336

Scoones I (1998) Sustainable rural livelihoods: a framework for analysis. IDS working paper 72. Available via IDS. https://www.ids.ac.uk/publications/sustainable-rural-livelihoods-a-framework-for-analysis/. Accessed 18 Mar 2019

Scoones I (2009) Livelihoods perspectives and rural development. J Peasant Stud 36(1):171–196

Smith W (1998) The rural linkages of urban households in Durban, South Africa. Environ Urban 10(1):77–88

Steinbrink M (2007) Exploring the role of migrants' networks in the context of translocal livelihoods in South Africa. In: Lohnert B (ed) Social networks: potential and constraints. Indications from South Africa. Verlag für Entwicklungspolitik, Saarbrücken, pp 73–113

Steinbrink M (2009a) Leben zwischen Land und Stadt, Migration, Translokalität und Verwundbarkeit in Südafrika. Springer, Wiesbaden

Steinbrink M (2009b) Urbanisation, poverty and translocality insights from South Africa. Afr Popul Stud 23:219–252

Steinbrink M (2012) Migration, Netzwerk und, Entwicklung'. Translokale Livelihoods in Südafrika. IMIS-Beiträge 42:165–204
Tacoli C (1998) Rural-urban interactions: a guide to the literature. Environ Urban 10(1):147–166
Thieme S (2008) Sustaining livelihoods in multilocal settings: Possible theoretical linkages between livelihoods and transnational migration research. Mobilities 3:51–71
van Dijk T (2011) Livelihoods, capitals and livelihood trajectories: a more sociological conceptualization. Prog Dev Stud 11:101–117
Vorlaufer K (1992) Wanderungen zwischen ländlichen Peripherie- und großstädtischen Zentralräumen. Zeitschrift für Wirtschaftsgeographie 28(4):229–261

Chapter 4
Vulnerability and Translocality: Why Livelihoods Become Translocal

Abstract In this chapter, we further elaborate the concept of translocal livelihoods to establish a basis for answering the following questions: Why does the translocalization of livelihood systems in Africa occur? And to what extent does the translocalization of livelihoods relate to different degrees of vulnerability? A *model of translocality* is introduced. Based on Anthony Giddens' idea of "structuration", it enables us to analyze translocal action and scrutinize the significance of such action for the structuration of the translocal. The model thus provides an analytical framework for answering the questions of (a) how different aspects of translocal strategic action contribute to the formation and reproduction of translocal structures and (b) how translocal social linkages affect the embedded actors' livelihood strategies (Sects. 4.1 and 4.2). In the Sects. 4.2.1, 4.2.2 and 4.2.3 three important types of strategies—*economic diversification, migration* and *social networking*—are each be examined in terms of vulnerability and translocality.

Keywords Translocal structuration · Translocal economic diversification · Transmigration · Translocal social networks

In this chapter, we further elaborate the concept of translocal livelihoods to establish a basis for answering the following questions: Why does the translocalization of livelihood systems in Africa occur? And to what extent do the translocalization and translocality of livelihoods relate to different degrees of vulnerability?

M. Steinbrink and H. Niedenführ, *Africa on the Move*, Springer Geography,
https://doi.org/10.1007/978-3-030-22841-5_4

4.1 Translocal Structuration: Translocality as Consequence and Condition of Livelihood Agency

In order to better understand the informal, translocal interlinkages in Africa and their feedback effects on processes of rural structural change and regional development there, and to grasp them with the concept *translocal livelihoods*, we first need to consider the process-like character of space-spanning structures of interconnectedness. When analyzing translocal socio-spatial structures, it is useful to take the dialectical relation of *translocal agency and translocal structure* of social interlinkages into account.

Here Anthony Giddens' concept of structuration (1979, 1984), which refers to the fundamental "*duality of action and structure*", is helpful.

In loose reference to Giddens (1992, 1997), Steinbrink (2009a, b: 90f) writes of the "*duality of the translocal*" in his "*small theory of the structuration of the translocal.*" According to this idea, the relationship of translocal agency and translocal structure does not refer to two independently given sets of phenomena—a dualism; instead, both elements together constitute a *duality*. Consequently, the "*duality of the translocal*" indicates that the structural elements of translocal social interweaving are both the medium and outcome of the translocal practices which recursively produce the socio-spatial linkages. Borrowing again from Giddens (1979, 1984), this can be called the "*structuration of the translocal.*"

Structuration of the translocal: *translocal interdependencies as medium and result of translocal practices.*

To conceptualize this *translocal structuration*, we suggest first differentiating between (a) *translocalization* as a process and thus translocality as a consequence of action and (b) *translocality* as actual state and as a condition for action.

On (a) Translocality as a Consequence of Action

Translocal structures are always the (temporary) result of a continual process.

Translocal socio-spatial structures in Africa are not to be taken as given; translocality emerges, sustains itself and changes over time. In this respect, existing translocal structures must principally be interpreted as the (temporary) results of a continual process.

Translocalization is the spatial expression of livelihood adaptation.

In the view of livelihood and vulnerability research, this process is directly related to the *translocalization of livelihoods*. Livelihood systems, for their part, are not static but constantly subject to a regulated dynamic of change.

Livelihood systems are always the results of a more or less successful adaptation of livelihood strategies in reaction to changing contexts of action. Translocalization and the resulting translocal interconnectedness are, consequently, the spatial expression of this adaptation process, and thus translocal socio-spatial structures can be read as a consequence of changing livelihood strategies.

On (b) Translocality as a Condition for Agency

Translocal structures are an essential element of the action context to which the actors must adapt.

Translocal interconnections are not only an outcome of action and a consequence of the translocalization of livelihood systems; they also constitute a crucial aspect of the context of action, which impacts the decisions of the embedded actors.

The social linkages that develop in the course of the translocalization of livelihoods result in structures that invariably have an enabling as well as constraining character. Thus—on the meso-level of social networks, so to speak—translocal structures form an important aspect of the context of action. It is to these translocal realities that translocal households must adjust, continually adapting their objective setting and the respective livelihood strategies. In this way, we can say that the adaptation of livelihood strategies takes place within translocal structures; in these structures, strategic action is taken and through that action the translocal socio-spatial structures are recursively reproduced as consequences and conditions of action.

Understanding the "duality of the translocal" is centrally important when analyzing structures of translocal social interweavement, because it demonstrates that conceptualizing such structures as independent of agency is futile. Further, this understanding makes clear that translocality does not exist in and of itself but can only be grasped as a recursive practice. This affects our understanding in two way: First, we must view translocal structures as consequences of rational, translocal action. Second, we must consider translocal actions (migrations and space-spanning social interactions) as being embedded in translocal structures, meaning they should be explained both within and based on this context.

4.2 Translocalization in the Context of Social Vulnerability

People act differently in contexts of risk and uncertainty than under conditions of relative safety.

By no means should we assume that only people who are particularly exposed to existential risks and uncertainties migrate and build translocal structures. The translocalization process is not always an expression of high vulnerability. However, considering the precarious living conditions prevalent in many rural and urban areas in Sub-Saharan Africa, we can assume that many of the actors involved in translocal contexts can be classified as socially vulnerable. In this respect, when considering translocality in Africa, we should explicitly consider the particularities of agency in the context of vulnerability, because people undoubtedly act differently in the face of risk and uncertainty than under conditions of relative security. Not only do vulnerable actors, obviously, generally lack certain means (resources, assets, etc.) for achieving the goals they set; what is more, the degree of vulnerability they experience influences their fundamental orientation for action.

The maxim of benefit/profit maximization is not universal but merely one form of action orientation and, as such, not rational per se

At times, we must remind ourselves that the maxim of benefit/profit maximization is by no means universal but merely *one* possible form of action orientation; nor is it rational per se. Action aimed at maximizing benefits can only be regarded as rational if no existentially threatening risks are to be expected. In such cases, benefit maximization actually reflects a freedom of action. Conversely, the greater the individual uncertainty, the more irrational an orientation along the maxim of maximization would be.

The rationality of safety

Clearly, at least one other rationality of action orientation must exist in addition to the principle of benefit/profit maximization: *the pursuit of safety*.[1] James C. Scott was one of the first scholars to draw attention to the *rationality of safety*. In his influential work on the "Moral Economy of the Peasant" (Scott 1976), he elaborates the "*safety-first principle*." The following quotation exemplifies Scott's conceptual starting point:

[1]For a detailed discussion on "action-centered simplifications" in vulnerability research, as well as on "context-variant action logic" and the "rationality of objective setting" ("*Rationalität der Zwecksetzung*") in the context of social vulnerability, see Steinbrink (2009a, b: 68–82).

> It is perfectly reasonable that the peasant who each season courts hunger and all its consequences should hold a somewhat different opinion of risk-taking than the investor who is gambling 'off the top'. [...] Given the social reality of the subsistence crises level for most peasant cultivators, it makes eminent sense for them to follow [...] the safety-first principle. (Scott 1976: 15)[2]

This shows that there is no single, universal logic of economic action. Specifically, in addition to the rationality of wealth maximization, we can discern the *rationality of risk minimization*.

The survival rationality

Moreover, it is necessary to expand this differentiation in order to understand the livelihood strategies of those households already living in a state of complete impoverishment and in acute life-threatening situations: They orientate their action neither toward benefit maximization nor along the safety maxim but solely toward survival (Shahabuddin et al. 1986; Shahabuddin and Mestelman 1986). If the livelihood system breaks down—i.e. in disaster situations—action is no longer oriented toward the goal of restabilization; instead actors are compelled to follow a short-term *survival logic*. In such a situation, the objective is purely the preservation of life. This action logic can thus also be described as the *rationality of desperation* or the *rationality of survival*.

Depending on the degree of vulnerability, we can thus distinguish *three basic logics of action orientation*:

(1) Rationality of benefit/profit/wealth maximization (*maximization rationality*)
(2) Rationality of risk minimization (*safety rationality*)
(3) Survival rationality (*rationality of desperation*).

Figure 4.1 illustrates the context-specific relationship between the degree of vulnerability and these rationalities of objective setting.

---

[2]Scott (1976) deduces this "safety-first principle" in the context of smallholder subsistence agriculture. Considering the minimum harvest yield necessary to meet the subsistence level needs of a household and the variations in actual harvest yields, he concludes that the risk aversion frequently observed in economic decision-making in smallholder contexts reflects a very rational economic logic—a logic which cannot be reconciled with the neoclassical principle of profit maximization, however: "In the choice of seeds and techniques in cultivation it means simply that the cultivator prefers to minimize the probability of having a disaster rather than maximizing his average return." (Scott 1976: 17f)

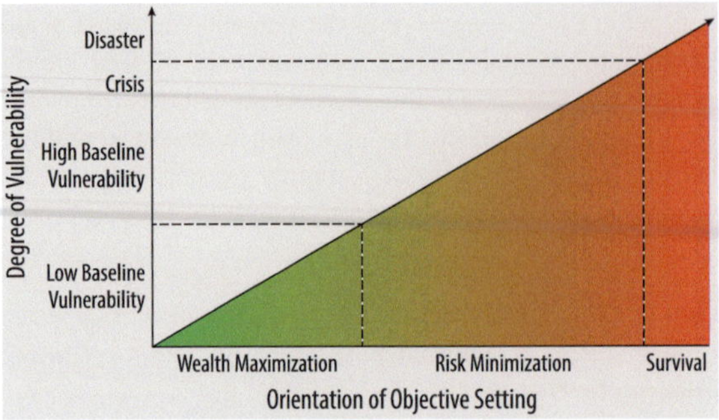

Fig. 4.1 Context-specific rationality—vulnerability and the orientation of objective setting. *Source* Own illustration

This differentiation among context-specific action orientations enables a more nuanced approach to understanding (translocal) action and its consequences for rural development dynamics.

Economic, social and spatial diversification are means of minimizing risk.

Most research on social vulnerability addresses the "safety-first principle" and is based implicitly or explicitly on the assumption that vulnerable actors rationally and strategically orient their actions toward the objective of minimizing risks. An essential means of minimizing risk is *risk distribution*. Studies of livelihood and vulnerability research have primarily investigated two types of strategies for the dispersion of risk:

Economic risk distribution: This strategy type refers to economic diversification which combines different economic activities, broadening the household income portfolio to offset potential income risks (see e.g. Ellis 1998; Barrett et al. 2001).

Social risk distribution: Here, this strategy type does not refer to the spreading of social risks but to the spreading of economic risks across social networks in which mechanisms of redistribution are at work, thus serving as an informal social security system for the embedded actors.[3]

[3]The emergence of the vulnerability approach at the end of the 1980s was a reaction to the purely economic approaches previously dominant within development research, as well as to conventional concepts of poverty which were limited to monetary income variables. Vulnerability research contradicted the notion that "the poor" represented a homogeneous and passive mass (see Chambers 1989; Swift 1989). Accordingly, vulnerability was no longer equated with material poverty but interpreted as a multidimensional and context-specific situation of

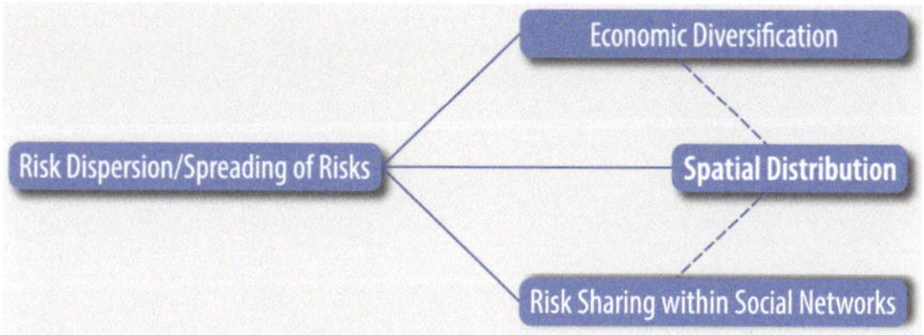

Fig. 4.2 Forms of risk dispersion. *Source* Own illustration

However, researchers have primarily analyzed both social networks and economic diversification in the *local* context. Often, too little consideration has been given to a third type of strategy for risk minimization: risk distribution through spatial diversification (*spatial risk distribution*) (Fig. 4.2).[4]

Central aspects of translocal livelihoods:

1. *translocal economic diversification*
2. establishment and maintenance of translocal networks
3. *transmigration*

In order to understand how the translocal fabric of interconnection emerges and persists in Africa, we need to recognize these three types of risk-minimizing strategies as essential aspects of translocal livelihoods. Primarily, we must consider (1) *translocal* economic diversification, (2) the establishment, maintenance and utilization of informal, *translocal* social networks, and (3) *trans*migration.

These three aspects are closely interrelated and cannot be analyzed independently, because they are interlocked in their relations as prerequisites and conditions. The interrelation of these three aspects is as obvious as it is revelatory for the analysis of the translocal. Figure 4.3 illustrates this idea graphically.

This basic model of translocality (see Fig. 4.3) serves as a conceptual starting point for the analysis of translocal

---

deprivation, in which those involved develop strategies and take intentional action. The discussion has always recognized *social isolation* as a factor influencing vulnerability (see Chambers 1989; Chambers and Conway 1992; Racodi 1995; Bohle and Watts 1993; Hufschmid 2011). In this context, social networks are understood as informal social safety nets, a *coping strategy* that can be called upon in times of crisis. Since the 1980s, the study of social networks and their significance for securing the survival of marginalized population segments has steadily gained popularity within development research. And the concept of *social capital* as a resource of "the poor" has become popular (see Steinbrink 2012).

[4]Although some vulnerability studies have investigated migration, they have mostly viewed it as a *coping strategy* in times of acute crisis or disaster. Little attention has been paid to the significance of migration as planned strategy in "normal times" and an integral part of the livelihood system.

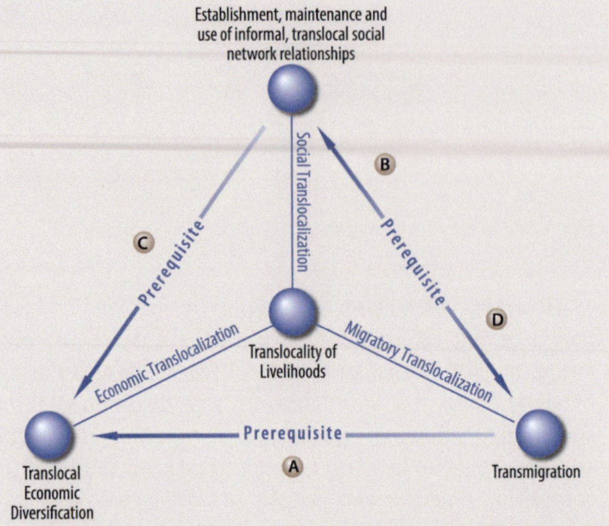

Ⓐ **Economic diversification** is increasingly achieved by combining different economic activities in different places ***(economic translocalization)***. A prerequisite for this is migratory translocalization, i.e. the migration of members of social and economic units who remain part of those units even after migrating. These migrants are referred to here as **transmigrants** and the social units translocalizing in the course of transmigration are referred to as *translocal households* (see above).

Ⓑ Through transmigration, social networks emerge that cross territorial demarcations and stretch between the region of origin and arrival ***(social translocalization/translocality)***. The process of social translocalization results in the spatial expansion of social network structures. **Translocal social networks** persist as long as translocal social action takes place within these structures.

Ⓒ Translocal social ties form the basis for the relationships of cooperation and exchange which are indispensable for **translocal economic diversification**, i.e. the combination of economic strategies at different places. Social translocalization is thus the prerequisite for ***economic translocalization/translocality***.

Ⓓ Furthermore, existing **translocal social network** relations with people at a potential destination are an essential condition for further migration, as they enable, among other things, access to the labor market and housing opportunities for potential migrants, thus reducing migration costs and risks. This can lead to chain migrations and induces dynamics which can be referred to as ***cumulative causation of translocalization***.

Fig. 4.3 Analytical model of translocality. *Source* Own illustration

socio-spatial structures. It enables us to analyze translocal action as strategic agency and scrutinize the significance of such action for the structuration of the translocal. The model thus provides an analytical framework for answering the questions of how the different aspects of translocal strategic action contribute to the formation and reproduction of translocal structures and how translocal social linkages affect the livelihood strategies of the embedded actors.

In the following chapters, the three strategy types—economic diversification, migration and social network(ing)—will each be examined in terms of vulnerability and translocality.

4.2.1 Economic Diversification, Vulnerability and Translocality

"Diversification is the norm."

Africa is still perceived as a continent of subsistence farmers; however, most households—even in rural areas of Africa—do have non-agricultural sources of income, albeit often on an extremely modest scale. Today, a relatively large proportion of the average monetary household income is generated outside of the agricultural sector.[5] That means: "Diversification is the norm" (Barrett et al. 2001: 1).[6]

The phenomenon of economic diversification has attracted increasing attention in research since the 1980s. However, this research has clearly concentrated on rural areas, meaning that diversification is usually examined in the context of rural development and rural socioeconomic change.[7] In livelihood

---

[5]In a comparative study Davis et al. (2017: 157) state: "Turning to income shares, the countries in our African sample show a tendency toward on-farm sources of income (i.e. agricultural income minus agricultural wages): they have higher shares of on-farm income (63%) and lower shares of non-farm wage income (8%), compared with countries of other regions (33 and 21% respectively), including those at similar levels of GDP. All the countries from Sub-Saharan Africa in this sample earn at least 55% of their income from agricultural sources, reaching approximately 80% in a number of countries (Ethiopia, Madagascar, Malawi, and Nigeria in 2004). Similarly, on-farm income accounts for more than 50% in all but one country (Kenya, at 48%). Combined with the observation above on the virtually universal level of participation in agricultural activities in the Sub-Saharan Africa subsample, this reinforces the message of agriculture still dominating the rural economy on the continent. Despite the fact that non-agricultural activities are ubiquitous (70% participation), they still account on average for only about one third of total earnings."

[6]See Bryceson and Jamal (1997), Reardon (1997), Little et al. (2001), Barret et al. (2001), Bryceson (2002), Ersado (2006: 3), Davis et al. (2017: 153).

[7]Since the late 1990s, however, it has been well documented that diversification is not a purely rural phenomenon but also an important strategy for large sections of the urban population in developing countries (see Rakodi 1995; De Haan 1999; Moser 1998).

and vulnerability research in particular, the concept of economic diversification has been very prominent right from the start.[8] At the heart of this discussion is the strategic aspect, i.e. the importance of diversification for livelihood security and as a potential way out of poverty and vulnerability.[9] In accordance with the livelihood research paradigm, the household is the central unit of consideration; consequently, the focus lies on the combination of household members' various economic activities.[10]

Despite the long and extensive discussion in development research about economic diversification in Africa, the debate does not provide any clear answers to the crucial questions of the cause and effect of diversification (Ellis 1998): In some studies, diversification is seen as an expression of actors' freedom of choice and action and as a seizing of existing economic opportunities (see e.g. Stark 1991), while in others it is seen primarily as an involuntary reaction to crisis (e.g. Davies 1996). Sometimes diversification is praised as an opportunity for the poor to gain more economic security (e.g. Zoomers and Kleinpenning 1996), while other studies view economic diversification as a special option of the wealthy that allows them to further increase their affluence (e.g. Hart 1994).

The effects of diversification on smallholder production systems have also received disparate assessments within the literature. Some authors try to demonstrate that the agricultural productivity of smallscale farms increases as a result of the investment capital acquired outside of agriculture (e.g. Stark and Bloom 1985; Carter 1997). Other empirical studies

---

[8]" […] diversification lies at the heart of livelihood strategies in rural Africa" (Barrett et al. 2001: 1).

[9]The combination of different economic strategies is described in the literature as "occupational multiplicity" (see White 1976; Breman 1996), "multi-tasking" (see Zoomers 2001), "income diversification" (see Barrett et al. 2001) or "livelihood diversification" (see Ellis 1998). According to the various research interests in this area, different criteria are used to categorize the combinations of economic activities. Among these categories are: agricultural/non-agricultural, informal/formal, wage labor/self-employment and local/migratory (see Ellis 1998:5 and Barett et al. 2001: 6ff).

[10]Diversification can take place (a) within one activity (e.g. simultaneous agricultural production for subsistence and market purposes), (b) through the combination of various economic activities carried out by one person or (c) through the combination of different activities carried out by several household members.

conclude that smallscale farms suffer considerably as a result of diversification into non-agricultural areas, as necessary resources—above all workforce—are diverted from farming (e.g. Low 1986; Lipton 1977; Preston 1989). According to Bryceson (1996, 2000, 2002), diversification is decisive in the process of deagrarianization taking place in many parts of Africa, i.e. the steadily declining importance of agriculture in rural areas.

Further, we can discern different lines of argumentation concerning the reasons for diversifying economic activities. On one hand, the more economically-oriented literature cites the shortcomings of land, labor or credit markets (see e.g. Barrett 1997; Reardon et al. 2000; Little et al. 2001; Barrett et al. 2001). On the other hand, agrosociological studies often refer to agricultural seasonality as the key explanation:

> Seasonality on its own explains many of the patterns of diversity in rural household incomes […]. In economic terms, seasonality means that returns to labour in both on-farm activities and off-farm labour markets vary during the year, causing seasonal changes in occupation as labour time is switched from lower to higher return activities. (Ellis 1998: 11)

Economic diversification as risk minimization: In favor of lower income variability, actors forego a potentially higher income by combining economic activities with the lowest possible covariance in terms of income risk.

However, one common feature found repeatedly in arguments on the reasons for economic diversification is risk minimization. Favoring a lower income variability, actors forego potentially higher incomes and instead combine economic activities with the lowest possible covariance in terms of income risk. In accordance with the *safety-first principle* (see above), households do not expand their income portfolio in order to make a profit, but rather based on rational safety considerations.[11] Accordingly, the majority of the studies on economic diversification in rural areas of Africa conclude that the process of income portfolio diversification for rural households ought to be understood as the adaptation of livelihood systems to a risky environment (*risk management strategy*).

Economic diversity of livelihood as a symptom of vulnerability?

Clearly, it would be rash to assess economic diversification all too optimistically as *the* strategy for the poor to reduce their vulnerability. For one thing, safety orientation often reflects *objective setting* in the face of precarious life

[11]See Alderman and Paxson (1992), Bryceson (1996), Dercan and Krishnam (1996), Scoones (1998), Ellis (1998), Francis (2000) as well as Barrett et al. (2001).

contexts and is frequently a reaction to acutely deteriorating conditions for action. Looking at it from this perspective, an observable economic diversity of livelihoods is thus more likely to be a symptom of vulnerability than a way out of it. On the other hand, some case studies in rural regions in Africa point to a positive correlation between the degree of household diversification and various prosperity indicators[12] and assert that it is not (only) the poor and vulnerable who diversify. This observation stands in some contrast to the notion that risk minimization is the primary motive for diversification and suggests that economic diversification can also serve profit maximization. That means, if there are no existential risks threatening the household, accumulation can also be the motive for diversification.

The widely diverging results of research on the economic diversification of livelihoods in Africa indicate that there are no general answers to the questions of who has what opportunities for diversification, who is subject to which constraints and why. Economic diversification is an "infinitely heterogeneous social and economic process, obeying a myriad of pressures and possibilities" (Ellis 1998: 29). Decisions on economic diversification are always embedded in cultural, social and political-institutional contexts, which can vary greatly from place to place. Ultimately, causes and effects can only be investigated in individual, local and—as emphasized here—*translocal* studies.

The translocal perspective of livelihood research compels us to view economic diversification in its space-spanning context.

The translocal livelihood perspective requires us to view economic diversification in its space-spanning context. For livelihood diversification is by no means a local phenomenon; instead, it is increasingly evident in the combination of economic activities in different places (e.g. rural and urban locations). Many studies dealing with livelihood security in rural Africa point to the great importance of migration for the expansion of income portfolios and emphasize the importance of remittances for the income of the rural population (see De Haan 2006).

Economic translocalization—i.e. the combination of different economic strategies at different places—marks an important intersection between economic diversification and migration. On the one hand, a natural prerequisite for translocal economic diversification is the migration of household members who remain part of their social unit

[12]On this, see the overview in Reardon (1997).

despite spatial separation. These individuals ("movers") are still closely connected to those who remain behind ("stayers") through relationships of cooperation and exchange. On the other hand, the aim of diversifying (or need to diversify) is known to be a central motive for migration. And often the pursuit of economic diversification is what triggers translocalization processes.

To grasp the connection between economic diversification and the translocality of livelihood systems, the migrant must not be viewed in isolation: The majority of jobseeking migrants do not move in order to leave the household but instead to perform a "function" within the household collective, usually in accordance with specific role expectations. In this way, labor migration must be understood both as an individual act and as part of a collective household strategy, motivated primarily by the aim of diversification. Only against this backdrop can migration and phenomena like remittances be meaningfully incorporated into the discussion on the diversification of livelihoods.[13]

Economic translocalization is a form of adaptation; and this adaptation process results in multiple, translocal livelihoods.

As described above, the economic diversification efforts of households can be based on different action orientations, depending on the respective conditions for action. Logically, the same applies to *translocal* portfolio diversification. According to its context-dependent rationality of objective setting, economic translocalization can be either profit oriented (*rationality of maximization*) or security oriented (*rationality of security*)—or in extreme cases, when all members remaining would represent an existential risk, survival oriented (*rationality of desperation*).

So, whereas economic translocalization is a form of adaptation, that adaptation process results in multiple, translocal livelihoods. Translocal diversification can either be a short-term variation in the household's economic strategy

[13]The "new economics of labor migration" approach to migration studies is based on a similar perspective (see e.g. Stark 1984, 1991). It emphasizes the importance of families and households and their collective strategies as central elements of migration processes. While classical migration theories assume an autonomously acting individual, this approach puts families or households first. Focusing on the analytical unit of the household means that the household income is considered rather than the individual incomes of its members. Here, the aggregated benefit to the household may well be at odds with individual benefit. As a result, when examining migration decisions, researchers must also consider the expected overall household benefit or the overall risk to the household.

(*coping*)[14] or part of a longer-term, planned shift in livelihood strategies.

How the process of translocal economic diversification affects the vulnerability of a household in the long run depends on why which households translocalize in which action contexts. Whether translocal households simultaneous use of distinct opportunities in different places is an expression of compulsion or freedom of action, whether the actors' action logic is oriented toward survival, safety or profit—such questions can ultimately only be answered in concrete case studies. There are no generalized answers here either. However, it is plausible that the greater the freedom of action, the more likely it is that economic translocalization will be part of a positive adaptation process. Whereas the more translocalization is an expression of an existential indispensability of action, the more likely it is to have negative medium- or long-term consequences for the household's livelihood security and wellbeing.

4.2.2 Migration, Vulnerability and Translocality

Livelihood and vulnerability studies long regarded migration in Africa primarily as a form of flight: either as the culmination of progressively escalated state of vulnerability or as the reaction to an acute crisis or disaster ("rural exodus"). Migration was thus interpreted as an exception or an anomaly. This view corresponds with the still widespread assumption in migration and development research that sedentarism is the norm (Verne 2012).[15]

The approach of translocal livelihoods is directed against the notion of sedentarism as the *norm: Migration also takes place under "normal conditions"*

In light of the empirical fact that many migratory events in Africa are integral parts of livelihood security, it is time to abandon this notion of sedentarism as the universal norm. The translocal livelihoods approach offers an alternative view, highlighting how migration often facilitates the functioning of livelihood systems. The strategy of migration is not

[14]One example would be the short-term wage work of a household member in an urban center to compensate for an unexpected loss of income following a crop failure.

[15]"(…) Migration is perceived as a spatially and temporally very limited anomaly in an otherwise sedentary life" [translation] (Pries 2001: 8).

only utilized in extreme cases but often under "normal" circumstances; although these can themselves be strongly determined by risk, insecurity and the threat of crises, which all too often constitute "African normality."

The differentiation of several possible action orientations serves to more precisely categorize the context-dependent stimuli for migratory processes. As a rational act of survival, migration can be an expression of compulsion, but it can also reflect a relatively free decision-making process; the strategic aims of a migration act may be oriented toward survival, risk minimization or profit maximization.

On this basis, we can make a sounder assessment of the *effects of migration* (e.g. on processes of rural development). Positive or negative consequences are no longer derived from the prevailing set of presumptions but rather from an empirical analysis of the action context. In this way, we can account for the context dependence of migration. This allows us to plausibly infer that migrations based on a free decision are more likely to have positive effects than migrations based on a lack of viable alternatives.

The conceptualization of regions of origin and arrival as largely discrete social spheres, which has long predominated in migration studies (see Sect. 4.2.2), hinders our understanding of migration as a means of securing a livelihood:

> […] geographical categories are less important than understanding the role migration plays in livelihood strategies. In fact, geographical distinctions can limit our understanding of livelihoods. (De Haan 2000: 2)

In contrast, the translocal livelihood approach provides insight into the importance of migration for livelihood security. At the same time, it makes clear how significant the strategy of migration is for the socio-spatial fusion of regions of origin and target regions into a single translocal social space.

According to the idea of *translocal structuration* (see Sect. 4.1), migration is relevant to translocality in three ways: First, migration is a precondition for translocalization; second, it is an action that takes place within translocal social space; and with that, it is, third, an action that contributes to the reproduction of translocal networks as socio-spatial links.[16]

[16]The translocality of social space is a consequence and at the same time a condition of migratory actions, because the socio-spatial structure significantly influences the action contexts of the "movers" and "stayers" involved, and thus their action decisions.

The emergence of translocal migration contexts we call *migratory translocalization*. In accordance with transnationalism research, migration within translocal spaces—along translocal network ties—can be designated *transmigration*.[17]

In order to delineate the concept of transmigration, the following basic distinctions (types of migration) are helpful:

"(Socio-spatial) Emigration" means a one-time, unidirectional movement. This type of migration leads to a breaking with previous social ties.

- *(Socio-spatial) Emigration*: These are migrations which, in the sense of a one-time, unidirectional movement, signify leaving the previous social space and entering another. Here, migration is accompanied by a break with previous social ties. In the context of translocality, this type of migration plays only a marginal role, since it contributes neither to the emergence nor to the reproduction of translocal network structures.
  As noted above, most migration studies have been conceptually based on this type of migration. However, empirical realities in Africa show that only very few migrations take place outside of social networks.

"Expanders" migrate to areas not previously linked to the region of origin by translocal ties; existing social bonds are maintained.

- *(Socio-spatial) Expansion*: This type of migration is prerequisite for the translocalization of livelihoods.[18] The "expander" migrates to an area not previously linked to the place of origin by translocal ties and maintains his or her existing social bonds (through remittances, visits, telephone calls, etc.). Thus, he or she remains part of the household collective. Expansion usually takes the form of labor migration with the goal of economic diversification. In quantitative terms, expansionary migration probably accounts for only a small proportion of all migratory movements in Africa. Nevertheless, it plays a decisive role in initiating translocalization: As they depart their original social spaces, expanders draw the edges of these along with them, stretching them across territorial boundaries simply by maintaining intensive social relationships to the people in the areas of origin.[19] These relationships reduce migration risks and thus make other

[17]On the concept of *transmigration* in transnationalism research, see Pries (1997, 2002).

[18]In studies dealing with phenomena of chain migration and migration networks, this is usually referred to as *pioneer migration*. We prefer the term socio-spatial expansion, because it sounds less militaristic.

[19]The social capital inherent in these relationships co-migrates, so to speak, with the expanders and serves as an important, destination-specific resource for other potential migrants (see "translocalization of social capital" in Sect. 4.2.3).

people from the same area more likely to also migrate. In migration research, these self-sustaining or self-reinforcing processes are referred to as chain migration or the "*cumulative causation of migration*" (Massey et al. 1994). In this way, the expander forms the nucleus of an emergent translocal network, fulfilling the prerequisite for subsequent migration events to take place within the translocalized structures (see "*translocalization of social capital*" in Sect. 4.2.3).

"Transmigration" takes place within existing translocal structures. Previously established translocal networks are utilized in the migration process.

- *Transmigration*: These are migrations stimulated or directed by social contacts at the target location, which take place within existing translocal structures. Previously established translocal networks are utilized in the migration process; they constitute a determining factor in both the migration decision and the choice of destination. We are convinced that the majority of all migration in Africa takes place as transmigrations along these lines.

  In order to make transmigration more tangible as an element of translocal livelihood organization, a distinction must be made between two types of transmigration: (a) migration which leads to the translocalization of households, and (b) migration which takes place within already translocalized household structures.

  In the first case (a), social relations *outside* the household (other relatives, friends, etc.) are utilized for transmigration. We can assume that the majority of transmigrations of this type serve the purpose of economic diversification, meaning that it is predominantly a matter of labor migration.

  The second case (b) deals with migrations between the different locations of a translocally organized household. The special nature of social ties within translocal households means that, in addition to labor migration, other forms of migration (e.g., child migration, education migration, health migration, partner migration) are becoming increasingly common (see "*phase model*" in Chap. 5). The special character of intra-household relations enables the transmigration of economically dependent members.[20] Although these movements probably account for a sizable share of the total volume of

[20]The support claims made on members of the same household tend to be much greater and stronger than those made on members of the wider social circle. Household support is also based less on the prospect of a relatively short-term reciprocation than on a long-term strategy.

migration in Africa, they receive considerably less attention from social scientists and planners than labor migration. This is likely because these forms of migration do not easily fit into the models of traditional macro-structural or micro-economic approaches.

The translocalized livelihood perspective highlights a variety of motives for transmigration: These migratory movements between different household locations may facilitate economic diversification, but they may also be an essential step toward utilizing the resources and opportunities available at different locations (e.g. access to hospitals and educational services). In addition, the differential costs of living at each respective household location is also decisive for certain migrations. For example, because the cost of living in rural areas of Africa is normally much lower, remigration is frequently used as a strategy in cases of prolonged unemployment or disability due to illness or old age. Beyond that, migrations taking place within translocal households also often serve to maintain the translocal organization itself. This involves, for example, household members moving to secure property or tenure rights, as well as moves to take over domestic tasks, such as child-rearing or caring for the elderly or sick, at another household location. This last point reveals that translocal household structures not only make migrations possible, but also sometimes make them necessary.

Transmigration between the different household locations can facilitate economic diversification as well as the utilization of resources available at those other locations (e.g. access to education and healthcare services).

The translocal fabric of interconnectedness created, utilized and reproduced via *migratory translocalization* is thus by no means the result of a single universal logic of action. Instead, grounds for and types of migration vary considerably.[21] In this way, we can say that translocality comprises the totality of a great and diverse assortment of translocal strategies oriented toward different goals according to the various rationalities of objective setting, which reflect the specific situation of the individual household.

Translocality comprises the totality of diverse strategies for translocal action, which are oriented toward different goals according to the situation of the individual households involved.

[21]The postulate that the translocality of the social space is fundamentally an expression of actions to minimize economic risk falls short. Migratory actions to which there is hardly any viable alternative also contribute to migratory translocalization, as do migrations that are part of a long-term, planned strategy for adapting livelihood systems to maximize benefits.

Translocal livelihood organization allows the respective local conditions to coalesce into one translocal action context.

In Africa, migration (between rural and urban areas, for example) is certainly not an exceptional phenomenon, neither is it a one-off, unidirectional movement nor the individual strategy of autonomous actors. The truth is that migration is an essential element of collective livelihood strategy. Migratory translocalization leads to a spatial diversification of livelihoods that enables strategic engagement with the economic and non-economic opportunities, obstacles and crises at different interlinked locations. Through translocal livelihood organization, the respective local conditions coalesce into one translocal context of action for the actors involved.

4.2.3 Social Networks, Vulnerability and Translocality

That which the social sciences today call "network perspective" has its roots in various academic disciplines and strands of research (Steinbrink et al. 2013: 9–11). In research on social networks in Africa, social anthropology can look back on an especially long tradition. In development research, the work of the *Manchester School* in the Copperbelt (Northern Rhodesia/Zambia) is particularly well known (see Mitchell 1969). These studies focused principally on social networks in urban contexts and their transformations during the urbanization process.[22] The heart of these discussions was the significance of networks for mobilizing support in times of personal conflict, scarcity or crisis (see e.g. Wheeldon 1969; Lourenço-Lindell 2001). As early as 1961, Philip Mayer used elements of social network analysis to investigate the process of "urban adaptation" of *isiXhosa*-speaking rural-urban migrants in South Africa during Apartheid (Mayer 1961).

The concepts of *moral economy* and the "*economy of affection*" have also strongly influenced the network perspective in development research. They looked at economic activity in non-modern societies and the role of social and cultural institutions therein (see Scott 1976). The notion of moral economy emphasizes the paramount importance of collective security mechanisms, which guarantee the

[22]As the focus of interest in ethnology expanded from the rural to include urban contexts, researchers realized that, in addition to kinship, other types of relationships also play an important role.

members of pre-capitalist societies a subsistence minimum (see Booth 1994). Proponents of the economy of affection explain the exchange relationships in peasant societies primarily in terms of a social logic determined by cultural rules and moral imperatives (see Hyden 1983).[23]

Studies point to the relevance of social network relations during crises and describe how these links are intensified when conditions of uncertainty prevail.

Finally, the concept of social networks became even more popular in development studies as the *vulnerability approach* gained ground at the end of the 1980s. Here, social networks are understood as informal safety nets that could be a particularly important recourse during acute crises.[24] However, the assessments of their capacity to mitigate crises differ widely: Numerous studies point to the generally great importance of social network relations in times of crisis and describe how these links are intensified, especially when conditions of uncertainty prevail (e.g. Lohnert 1995; De Herdt and Marysse 1997).[25] Other authors stress the fragility of these informal social security systems in the face of repeated, severe crises (see, among others, Devereux 1999; Benda-Beckmann and Kirsch 1999). This issue was addressed relatively early with respect to food crises—and one of the key findings of these studies was that support relationships intensify in times of moderate scarcity but can collapse in phases of acute existential danger (e.g. in cases of disaster) (see for example Drèze and Sen 1989; Corbett 1988; Campbell 1990; Davies 1996; Adams 1993). By assuming that informal safety nets have only a finite capacity, these works represent an important counterweight to the coping and self-help euphoria of vulnerability research.

Within the *sustainable livelihood approach*, social network relations became even more central: In addition to *capabilities* and tangible assets, *intangible assets* were also highlighted as an important component of livelihoods. Within

---

[23]Critics have accused these studies of romanticizing the societies in question by ignoring the structural inequality in the distribution of power within them. For a discussion of the "economy of affection" in Africa, see Tsuruta (2008).

[24]See for example the contributions in Lourenço-Lindell (2001) and Simone (2001) and Miles (2001) in Tostensen et al. (2001).

[25]Lourenço-Lindell (2002: 30) notes that some authors even go so far as to regard scarcity and risk as preconditions for the development of support networks, citing the study by De la Rocha (1994) as an example. The studies by Stack (1974) and Lomnitz (1977) as well as the well-known works by Lewis (e.g. 1965) on the "culture of poverty" can be regarded as precursors of this idea. More recently: Small et al. (2010) discuss the nexus of culture and poverty.

Social networks are the result of a process; they are built and maintained —and exist only as long as they are actually used.

the well-known *livelihood framework*, such intangible assets are referred to as *social capital* (see below). Strikingly, social relationships are primarily seen as *resources*. In most studies, the network is regarded as an informal system that *exists* apart from formal security systems. Social relationships are thus often interpreted as *given*, which is why the aspect of *network building* is hardly addressed. However, networks are always the result of a process: They are created and maintained and exist only as long as they are actually used.

Until now, the discussion of social networks and vulnerability has paid too little attention to "*social networking*" as a strategic action. However, since social network relations are a crucial social resource for the individual seeking to achieve his/her goals, the establishment and maintenance of informal network relationships is an important objective of action.[26] It is about (1) strategic *recourse* to social networks and (2) strategic *building and maintenance* of social network relations ("*social networking*"):

(1) If—as shown above in Sect. 4.2—the degree of vulnerability determines one's action orientation, then it will also codetermine the aims of a *strategic recourse* to social relations. In previous vulnerability studies, the utilization of social resources has mostly been presented as a coping strategy in times of crisis, congruent with the safety rationality.[27] However, utilizing social network relationships is also a livelihood strategy in "normal times". Not only in periods of acute crisis or disaster do the livelihoods of large parts of the population in Africa depend extensively on their incorporation in social networks. Recourse to social network resources is part of everyday life. Still, given the conditions of vulnerability that prevail in many places on the African continent, the safety rationality certainly plays a decisive role; the goal of profit maximization only becomes relevant to social

[26]Etzold (2017: 53) points out that "Translocality is, however, not a given fact that automatically comes into being through migration. A translocal space rather has to be produced and reproduced actively through translocal practices."

[27]This also includes efforts not to overburden existing relationships with one-sided aid requests. In situations of immediate existential danger, however, such considerations take a back seat in favor of a pure survival rationality (see Sect. 4.2) with the result that, under certain circumstances, the breakdown of relationships becomes a risk one must take.

network relations as the degree of vulnerability decreases.

(2) Individuals do not innately possess social network relationships; on the contrary, networks intrinsically require continuous investment in "relationship maintenance". So, if the utilization of social resources as a coping and livelihood strategy is of existential importance, *social networking* is synonymous with creating a means of subsistence. The more important a social network is for securing one's livelihood, the stronger the efforts will be to build and expand it and to prevent it from falling apart. In keeping with the idea of context-variant action orientation, the degree of vulnerability thus also determines the action motives for strategic *social networking* —i.e. *investment in social capital*:

Building social capital is synonymous with creating a means of subsistence: Networking is a key livelihood strategy.

- With increasing degree of vulnerability, strategic social action will focus on building and maintaining those relationships that promise greater social security (risk minimization).
- Networking geared more toward accumulation will tend to target those social relationships considered beneficial in terms of benefit/profit maximization.
- In contrast, people whose livelihood systems have collapsed and who find themselves in life-threatening situations will hardly have the opportunity to strategically invest in social relationships. They are dependent on one-sided handout relationships and must take the risk of being dropped from the social safety net.

As such, when regarding *networking* strategy, we must bear in mind that the grounds for strategic networking vary greatly. Since the bases of social support networks cannot be reduced to a simple formula, what specific significance the informal social networks have for poverty reduction or the development of a particular group can only be determined on a case-by-case basis.

Within the translocal livelihoods approach, the concept of the social networks plays the paramount role. The translocal perspective is a network perspective, because the translocal fabric of interconnectedness is nothing more than a dense configuration of relatively durable social relationships between actors who reside in different places.

The translocal perspective is a network perspective: The space-spanning social networks are the constituent element of translocality; they are the consequence and the condition of livelihood actions.

According to the "duality of the translocal" (see Sect. 4.1), space-spanning social networks are understood as the constituent element of translocality and as the consequence and condition of action. In the context of livelihood research in Africa, the structure of translocal social networks is often the result of strategic action in the face of risk, insecurity or crisis, which is manifested in uni- or bilateral interactions and transactions between people at different places.

Social translocalization is the process of these network structures' spatial expansion transcending the boundaries of locality. The result of this process is *social translocality*. The latter is closely related to the two types of strategies discussed in Sect. 4.2.1 (*translocal economic diversification*) and Sect. 4.2.2 (*transmigration*). Expander migration (socio-spatial expansion) is the fundamental prerequisite for the social translocalization process, but subsequent migrations (transmigration) also intensify social translocalization, as the web of social interlinkages between the localities becomes increasingly dense. As Sect. 4.2.1 already established, the migration of expanders and subsequent transmigrants is often—but by no means exclusively—related to the strategy of translocal economic diversification, as this is one of the main motives for migration.

In order to analyze the processes of social translocalization, the concept "*translocal social capital*" is helpful. "Social capital"[28] is an apt metaphor for highlighting the social aspect of economic action as well as the economic

[28]The sociological concept of social capital is heavily criticized due to its instrumentalization by international organizations. The concept's use in developmental discourse is subject to well-founded critique: Ever since Loury (1977) introduced the term into social science, "social capital" has been gaining considerable ground. Yet, it cannot be regarded as a coherent or consistent concept, since, depending on the discipline, it is applied with varying underlying meanings. Owing to these conceptional uncertainties, scholars have repeatedly suggested applying social capital with great caution. Fine (1998) even goes as far as to assume a causal connection between the analytical uncertainties and the popularity of the "chaotic" social capital approach. In discussions of development, too, the notion of social capital has become a constituent element of discourse. Especially among the international donor organizations (such as the World Bank), social capital has become a leading concept. It is considered the missing link, which promises the resolution of very different developmental problems, ranging from "poverty" to "bad governance". Within the social capital discourse, the positive effects are over-emphasized and the potential of the networks as resources for the poor is overstated. Hence, structural pressure upon social networks and the constraints within them are literally programmatically neglected.

aspect of social action. The term is well suited to describe the resource character of social relations.[29] The concept is also valuable for explaining processes of translocalization: Though social capital is intangible, it is nonetheless relatively bound to space. One might say that social capital is located with those people between whom a strategically useful relationship exists. If these people are situated at the same place, we can speak of "*local social capital*"; in the other case, we can speak of "*translocal social capital.*" This simple differentiation proves helpful for describing the emergence of social network structures that transcend spatial boundaries.

As described above in Sect. 4.2.2, the dynamics of chain migration means that every migratory movement tends to increase the attractiveness of the destination area, since a support network develops there upon which migrants in need can rely. In this sense, the phenomenon can not only be regarded as a *cumulative causation of migration* (see Massey 1990) but as *cumulative causation of translocalization* (Steinbrink 2010). As summarized in Fig. 4.4, the translocality of livelihoods can be described as the outcome of a self-sustaining process of *translocalization and relocalization of social capital.*

As a structure, translocal networks (*social translocality*) are often a prerequisite for the feasibility of household strategies. This applies to both (a) social relations within and (b) relationships outside the translocal household:

(a) The translocal social ties *within households* are, by definition, a fundamental prerequisite for the organization of translocal households. Accordingly, these relationships of cooperation and exchange are also a necessary condition for strategically combining different economic activities translocally.
(b) The translocal social relations *outside the household* are also an essential social resource. If not for the utilization of the social capital inherent in these relationships (with friends, relatives, etc.), certain translocal strategies would not be possible. Sections 4.2.1 and 4.2.2 already discussed this with regard to diversification and migration.

[29]Hereafter, social capital is understood as an individual resource, according to Pierre Bourdieu (1986: 248): It is "the aggregate of the actual or potential resources which are linked to possession of a durable network of more or less institutionalized relationships of mutual acquaintance or recognition."

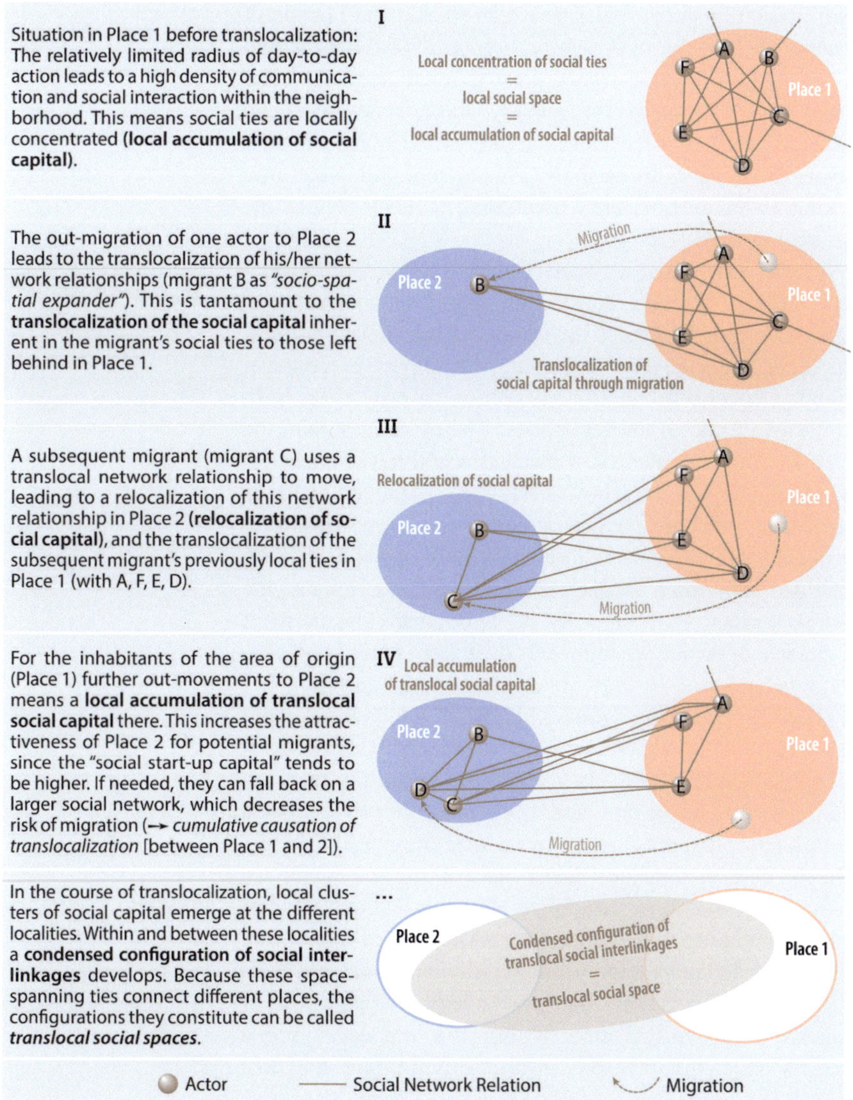

Fig. 4.4 The translocalization and relocalization of social capital as a cumulative process. *Source* Own illustration, modified from Steinbrink (2010: 51)

Informal, translocal social networks (inside and outside of households) link the opportunity structures of different places, and they often offer securities afforded neither by the market economy nor by the largely weak, state-run social welfare systems in Africa. Remaining in the translocal network is therefore a strong motive for action, and social networking is an indispensable livelihood strategy within the translocal context, too.

Aside from their enabling elements, social networks invariably have restrictive elements as well.

Nevertheless, social networks fundamentally comprise not only enabling but also constraining or even coercive elements. This is no different for translocal network structures. Ultimately, we can only determine the impact of concrete network mechanisms on strategic action case by case. Still, the following rule of thumb applies: The more important the social security provided by translocal social networks is for the individual—i.e. the more clearly safety-oriented action is emphasized—the greater his or her dependence within the translocal social context will be. And more dependent of the embedded individual is upon the network, the greater the network's sanctioning capacity is. This means the network has greater control and influence over the social (and economic) behavior of the individuals. Portes and Sensenbrenner (1993) describe this as "*enforceable trust*", which they regard as an important source of social capital. Hence, greater social network dependency entails additional constraints on individual freedom of action and may make negative effects (obstacles to prosperity) more likely (see Steinbrink 2012).

In the case of translocal social networking, we again see that it is impossible to generalize. The motivations for creating, maintaining and utilizing translocal social capital are also context dependent. Only for an individual case study can we assess which social mechanisms underpin the processes of exchange and redistribution within the translocal social space. Whether the intensive, informal rural-urban networks that exist in large parts of Africa are, in their respective manifestations, more an expression of relatively free action or a symptom of limited options for action in the context of vulnerability can ultimately be determined only for a specific case. This is another reason why it is extremely difficult to generalize about the extent to which integration into translocal structures is conducive or obstructive to actors' (or households') aims or what influence socio-spatial interaction has on local or regional contexts or even, possibly, on the development of society as a whole.

References

Adams A (1993) Food insecurity in Mali: exploring the role of moral economy. IDS Bull 24(4):41–51

Alderman CH, Paxson C (1992) Do the poor insure? A synthesis of the literature on risk and consumption in developing countries. Available via Documents-worldbank. Accessed 18 Mar 2019

Barrett CB (1997) Food marketing liberalization and trader entry. Evidence from Madagascar. World Dev 25(5):763–777

Barrett C, Reardon T, Webb P (2001) Nonfarm income diversification and household livelihood strategies in rural Africa: concepts dynamics and policy implications. Food Policy 26(4):315–331

Benda-Beckmann F von, Kirsch R (1999) Informal security systems in Southern Africa and approaches to strengthen them through policy measures. J Soc Dev Afr 14:21–38

Bohle HG, Watts MJ (1993) Hunger famine and the space of vulnerability. GeoJournal 30(2):117–125

Booth WJ (1994) On the idea of the moral economy. Am Polit Sci Rev 88:653–667

Bourdieu P (1986) The forms of capital. In: Richardson JG (ed) Handbook of theory and research for the sociology of education. Greenwood Press, New York, pp 241–258

Breman J (1996) Footloose labour: working in india's informal economy. Cambridge University Press, Cambridge

Bryceson DF (1996) De-agrarianisation and rural employment in Sub-Saharan Africa: a sectoral perspective. World Dev 24(1):97–111

Bryceson DF (2000) Disappearing peasantries? Rural labour redundancy in the Neo-liberal Era and beyond. In: Bryceson DF, Kay C, Mooij J (eds) Disappearing peasantries? Rural labour in Africa Asia and Latin America. IT Publication, London, pp 299–326

Bryceson DF (2002) Multiplex livelihoods in rural Africa: recasting the terms and conditions of gainful employment. J Mod Afr Stud 40 (1):1–28

Bryceson DF, Jamal V (1997) Farewell to farms: de-agrarianisation and employment in Africa. Ashgate, Aldershot

Campbell DJ (1990) Community-based strategies for coping with food scarcity: a role in African famine early-warning systems. GeoJournal 20:231–241

Carter MR (1997) Environment technology and the social articulation of risk in West African agriculture. Econ Dev Cult Change 45(3):557–591

Chambers R (1989) Vulnerability, coping and policy. IDS Bull 20(2):1–7

Chambers R, Conway G (1992) Sustainable rural livelihoods: practical concepts for the 21st century. Discussion paper 296, IDS Institute of Development Studies, Brighton

Corbett J (1988) Famine and household coping strategies. World Dev 16 (9):1099–1112

Davies S (1996) Adaptable livelihoods. Coping with food insecurity in the Malian Sahel. Palgrave Macmillan, London

Davis B, Di Giuseppe S, Zezza A (2017) Are African households (not) leaving agriculture? Patterns of households' income sources in rural Sub-Saharan Africa. Food Policy 67:153–174

De Haan A (1999) Livelihoods and poverty: the role of migration—a critical review of the migration literature. J Dev Stud 38(2):1–47
De Haan A (2000) Migrants, livelihoods, and rights: the relevance of migration in development policies. Social development working paper 4, London
De Haan A (2006) Migration in the development studies literature: Has it come out of its marginality? Research paper 19, United Nations University
De Herdt T, Marysse S (1997) Against all odds: coping with regress in Kinshasa, Zaïre. Eur J Dev Res 9(1):209–230
De La Rocha MG (1994) The resources of poverty: women and survival in a Mexican City. Blackwell, Cambridge
Dercan S, Krishnam P (1996) Income portfolios in rural Ethiopia and Tanzania choices and constraints. J Dev Stud 32(6):850–875
Devereux S (1999) Making less last longer: informal safety nets in Malawi. IDS discussion paper 373, Brighton
Drèze J, Sen A (1989) Hunger and public action. University Press, Oxford
Ellis F (1998) Survey article household strategies and rural livelihood diversification. J Dev Stud 35(1):1–38
Ersado L (2006) Income diversification in Zimbabwe: Welfare implications from urban and rural areas. World Bank policy research working paper 3964. Available via Worldbank. https://openknowledge.worldbank.org/bitstream/handle/10986/8389/wps3964.pdf?sequence=1&isAllowed=y. Accessed 16 Apr 2019
Etzold B (2017) Mobility, space and livelihood trajectories: new perspectives on migration, translocality and place-making for livelihood studies. In: De Haan L (ed) Livelihoods and development: new perspectives. Brill, Leiden, pp 44–68
Fine B (1998) The developmental state is dead–long live social capital. Dev Change 30:1–19
Francis E (2000) Rural livelihoods, institutions and vulnerability in South Africa. Paper presented at the DESTIN conference on new institutional theory, Institutional Reform and Poverty Reduction, London School of Economics, London, 7–8 Sept 2000
Giddens A (1979) Central problems in social theory: action, structure, and contradiction in social analysis. Univ. of California Press, Oakland
Giddens A (1984) The constitution of society: outline of the theory of structuration. Polity Press, Cambridge
Giddens A (1992) Die Konstitution der Gesellschaft. Campus-Verlag, Frankfurt a.M
Giddens A (1997) Die Konstitution der Gesellschaft: Grundzüge einer Theorie der Strukturierung, 3rd edn. Campus Verlag, Frankfurt a.M
Hart G (1994) The dynamics of diversification in an Asian rice region. In: Koppel B, Hawkins J, James W (eds) Development or deterioration: theories, experiences and policies. Oxford, New York
Hyden G (1983) No shortcuts to progress: African development management in perspective. Univ. of California Press, Berkeley
Lewis O (1965) La Vida: a Puerto Rican family in the culture of poverty—San Juan and New York. Random House, München
Lipton M (1977) Why poor people stay poor. A study on urban bias in world development. Harvard University Press, London

Little PD, Smith K, Cellarius BA, Coppock L, Barrett C (2001) Avoiding disaster: diversification and risk management among East African herders. Dev Change 32(3):387–419
Lohnert B (1995) Überleben am Rande der Stadt: Ernährungssicherungspolitik, Getreidehandel und verwundbare Gruppen in Mali; Das Beispiel Mopti. Verlag für Entwicklungspolitik, Saarbrücken
Lomnitz LA (1977) Networks and marginality: life in a Mexican Shantytown. Academic Press, New York
Lourenço-Lindell I (2001) Social networks and urban vulnerability to hunger. In: Tostensen A, Tvedten I, Vaa M (eds) Associational life in African cities: popular responses to the urban crisis. Nordiska Afrikainstitutet, Stockholm, pp 30–45
Lourenço-Lindell I (2002) Walking the tight rope informal livelihoods and social networks in a West African City. Acta Universitatis Stockholmiensis, Stockholm
Loury G (1977) A dynamic theory of racial income differences. In: Wallace PA, LeMund A (eds) Chapter of women, minorities, and employment discrimination. Lexington Books, Lexington, pp 153–186
Low A (1986) Agricultural development in Southern Africa: farm household theory and the food crisis. James Currey, London
Massey DS (1990) Social structure, household strategies, and the cumulative causation of migration. Population Index 56(1):3–26
Massey D, Arango J, Graeme H, Kouaouci A, Pellegrino A, Taylor E (1994) An evaluation of international migration theory: the North American case. Popul Dev Rev 4(20):699–751
Mayer P (1961) Townsmen or tribesmen conservativism and the process of urbanization in a South African City. Oxford University Press, Oxford
Miles M (2001) Women's groups and urban poverty: The Swaziland experience. In: Tostensen A, Tvedten I, Vaa M (eds) Associational life in African cities: popular responses to the urban crisis. Nordic Africa Institute, Stockholm, pp 64–72
Mitchell C (ed) (1969) Social networks in urban situations. University Press, Manchester
Moser C (1998) The asset vulnerability framework: reassessing urban poverty reduction strategies. World Dev 26(1):1–19
Portes A, Sensenbrenner J (1993) Embeddedness and immigration: notes on the social determinants of economic action. Am J Sociol 98 (6):1320–1350
Preston D (1989) Too busy to farm: under-utilisation of farm land in Central Java. J Dev Stud 26(1):43–57
Pries L (1997) Neue Migration im transnationalen Raum. In: Pries L (ed) Transnationale Migration. Soziale Welt. Sonderband 12:15–44
Pries L (2001) Internationale Migrationen. Soziologische Themen, Transcript, Bielefeld
Pries L (2002) Transnationalisierung der sozialen Welt? Berliner Journal für Soziologie 12(2):263–272
Rakodi C (1995) The household strategies of the urban poor: coping with poverty and recession in Gweru, Zimbabwe. Habitat Int 19:447–471

Reardon T (1997) Using evidence of household income diversification to inform study of the non-farm labour market in Africa. World Dev 25(5):735–748
Reardon T, Stamoulis K, Lanjouw P, Balisacan A (2000) Effects of non-farm employment on rural income inequality in developing countries: an investment perspective. J Agric Econ 51(2):266–288
Scott JC (1976) The moral economy of the peasant rebellion and subsistence in Southeast Asia: rebellion and subsistence in South East Asia. Yale University Press, New Haven/London
Shahabuddin Q, Mestelman S (1986) Uncertainty and disaster-avoidance behaviour in peasant farming. J Dev Stud 22 (2):740–751
Shahabuddin Q, Mestelman S, Feeny D (1986) Peasants behaviour towards risk and socio-economic and structural characteristics of farm households in Bangladesh. Oxford Economic Papers Nr. 38
Simone A (2001) Between Ghetto and Globe: remaking urban life in Africa. In: Tostensen A, Tvedten I, Vaa M (eds) Associational life in African cities: popular responses to the urban crisis. Nordic Africa Institute, Uppsala, pp 46–63
Small ML, Harding DJ, Lamont M (2010) Reconsidering culture and poverty. Ann Am Acad Polit Soc Sci 629(1):6–27
Stack CB (1974) All our kin: strategies for survival in a black community. Basic books, New York
Stark O (1984) Discontinuity and the theory of international migration. Kyklos 37(2):206–220
Stark O (1991) The migration of labour. Blackwell, Cambridge
Stark O, Bloom D (1985) The New Economics of Labour Migration. Am Econ Rev 75:173–178
Steinbrink M (2009a) Leben zwischen Land und Stadt, Migration, Translokalität und Verwundbarkeit in Südafrika. Springer, Wiesbaden
Steinbrink M (2009b) Urbanisation, poverty and translocality insights from South Africa. Afr Popul Stud 23:219–252
Steinbrink M (2010) The role of Amateur football in circular migration systems in South Africa. Afr Spectr 45(2):35–60
Steinbrink M (2012) Migration, Netzwerk und, Entwicklung‘. Translokale Livelihoods in Südafrika. IMIS-Beiträge 42:165–204
Steinbrink M, Schmidt JB, Aufenvenne P (2013) Soziale Netzwerkanalyse für HumangeographInnen: Einführung in UCINET und NetDraw in fünf Schritten (Vol. 5). Universitätsverlag, Potsdam
Swift J (1989) Why are rural people vulnerable to famine? IDS Bull 20 (2):8–15
Tostensen A, Tvedten I, Vaa M (eds) (2001) Associational life in African cities: popular responses to the urban crisis. Nordiska Afrikainstitutet, Stockholm
Tsuruta T (2008) Between moral economy and economy of affection. In: Maghimbi S et al (eds) Contemporary perspectives on African moral economy. Dar es Salaam Univ. Press, Dar es Salaam, pp 35–52
Verne J (2012) Living translocality: space. Culture and economy in contemporary Swahili trading connections. Steiner, Stuttgart
Wheeldon P (1969) The operation of voluntary associations and personal networks in the political process of an interethnic community. In: Mitchell JC (ed) Social networks in urban situations:

analyses of personal relationships in Central African towns. University Press, Manchester
White H (1976) Social structure from multiple networks. I. Blockmodels of roles and positions. Am J Sociol 81(4):730–780
Zoomers A (2001) Land and sustainable livelihood in Latin America. KIT Publishers, Amsterdam
Zoomers A, Kleinpenning J (1996) Livelihood and urban-rural relations in Central Paraguay. Tijdschrift voor Economische en Sociale Geografie 87(2):161–174

Chapter 5
Translocal Livelihoods in Sub-Saharan Africa

Abstract Having established the conceptual basis of the translocal perspective in the preceding chapters, this chapter focuses concretely on the phenomenon of translocal livelihoods in Sub-Saharan Africa. After a brief description of the specific characteristics of the African urbanization process (Sect. 5.1), we undertake a quantitative assessment of the significance of the translocal livelihoods on the African continent (Sect. 5.2). In Sect. 5.3, we then discuss some basic spatial and temporal patterns of migration and translocalization in Africa. And finally, we outline an ideal-typical "phase model of translocalization" (Sect. 5.3.3) and sketch out some social patterns within translocal livelihood systems by answering the questions: "Who stays, who leaves, who returns?"

Keyword Urbanization in Sub-Saharan Africa · Translocal livelihoods in Sub-Saharan Africa · Patterns of translocalization · Phase model of translocalization

This chapter focuses concretely on the phenomenon of translocal livelihoods in Sub-Saharan Africa. After a brief description of the specific characteristics of the African urbanization process and the role of translocalization in it (Sect. 5.1), we undertake a quantitative assessment of the significance of the phenomenon in Africa south of the Sahara (Sect. 5.2). Then, following the introduction of an ideal-typical "*phase model of translocalization*", we discuss some basic spatial, temporal and social patterns of translocalization in Sect. 5.3.

M. Steinbrink and H. Niedenführ, *Africa on the Move*, Springer Geography,
https://doi.org/10.1007/978-3-030-22841-5_5

5.1 Urbanization in Africa: "Urbanization Without Growth But with Translocalization"

Africa—the least urbanized yet fastest urbanizing continent.

The urbanization process is certainly one of the most prevalent spatial-structural dynamics on the African continent. Although, with an average degree of urbanization of about 40% in 2015,[1] (UN 2018: 10), Africa is still the least urbanized part of the world, the immense pace of urbanization poses far-reaching challenges. Because with an estimated annual urbanization rate of 3.4%[2] and a projected degree of urbanization of 56% in 2050, Africa is by far the fastest urbanizing continent (Beguy et al. 2010: 551–552).[3] Saghir and Santoro (2018: 1) point out that Sub-Saharan Africa's urban population of 472 million will double in the next 25 years. In addition to rural–urban migration, this vast urban expansion is fueled by strong natural population growth in urban areas.[4]

In contrast to North America and Europe, African urbanization is seldom linked to positive macroeconomic development.

However, the challenge lies less in the quantitative scale and intensity of the urbanization dynamics than in the economic and social conditions under which this enormous urban growth is unfolding. In contrast to North America and Europe, African urbanization is not tied to positive macroeconomic development; instead, it persists in periods of economic stagnation and in the face of immense national debt.

[1]For comparison: Europe 73%, world average 50%. In terms of the degree of urbanization, however, the major regions of Africa prove to be quite heterogeneous. In 2010, only 24% of the population in East Africa lived in cities, while in Southern Africa the figure was 59%. Central and West Africa had shares of 43% and 45%, respectively (UN Population Fund, State of World Population 2010: 105).

[2]For comparison: Latin America and Caribbean 1.6%; Asia 2.3%; North America 1.3%; Europe 0.4%; world average 1.9%. But the urbanization rates in the various major regions of Africa also differ strikingly: East Africa 3.8%; Central Africa 4.1%; North Africa 2.5%; Southern Africa 1.9%; West Africa 3.9% (UN Population Fund, State of World Population 2010: 105).

[3]"Urbanization projections indicate a slowing rate in Africa from the 1990s high rates of up to 8% down to a range of 1.9−2.2% from 2020 to 2050, with significant variation across its countries." (Güneralp et al. 2018: 1)

[4]According to some authors (Bocquier 2003; Bocquier and Traoré 2000), urban growth in Africa today is more strongly related to natural population growth than to rural–urban migration (Beguy et al. 2010: 551–552; Güneralp et al. 2018: 1).

Up into the 1960s, the phenomenon of urbanization in Africa seemed closely linked with the expansion of nonagricultural economic practices and growing productivity differences between rural and urban economies. For a long time, urbanity was synonymous with industrialization, modernity, progress, growth, prosperity, and development. Accordingly, differences in the degree of urbanization among individual states were thought to reflect different "stages of development" and a structural change in society as a whole (see Sect. 2.2). In light of the rapid urbanization in Africa, however, the postulated connection between increasing economic prosperity and urbanization has lost its empirical basis and has even been inverted: According to the UN Population Division (2015), the following countries have the highest urbanization rates in Africa: Rwanda (6.43%), Burkina Faso (5.87%), Burundi (5.66%), Uganda (5.43%), Tanzania (5.36%), Niger (5.14%), Eritrea (5.11%), Mali (5.08%), and South Sudan (5.05%). Six of these nine countries are among the ten African countries with the lowest HDI scores (see UNDP 2014: 159). This illustrates that it is now often the poorest countries that have the highest urbanization rates.

Often, it is the poorest countries that have the highest urbanization rates.

The reasons for the weak economic development in Sub-Saharan Africa since independence are manifold and complex. Suffice it to say that the newly founded African states, whose economies were still predominantly based on the export of raw materials and agricultural products, had a difficult starting position. One reason for this, among others, were the low prices on the world market. In addition to this crisis in export industry, the opportunities for (semi-)subsistence farming, which remains to this day essential for a large part of the rural population, have been progressively declining. Catalyzing this process are the strong natural population growth, the accelerating ecological degradation of large swaths of agricultural land and the competitive displacement of small-scale farmers by export-oriented commercial agriculture (*cash-crop production*).

Sub-Saharan Africa's economic crisis was first and foremost a rural crisis.

The economic crisis in Sub-Saharan Africa began primarily as a rural crisis, which became evident in a steadily declining rate of agricultural production per capita (see van Westen 1995). Strict government price regulations aimed at keeping food prices low for the urban population, in order to avoid food riots and social unrest, further exacerbated the situation of small-scale farmers. These policies reduced small-scale farmers' readiness to improve agricultural production or market products, which in turn had negative macroeconomic effects. The economic situation in many rural areas of Africa deteriorated severely, leaving fewer and fewer people able to secure their livelihoods without access to other monetary income sources. In many cases, their only recourse was to migrate to urban areas to seek work (see Gilbert 1992). On top of this crisis in small-scale agriculture, droughts, floods, and wars have had devastating, at times catastrophic effects: people have often had no choice but to leave their rural homes and move to the cities, which were frequently a last refuge.

Yet, largely due to their poorly developed production structures, Africa's cities do not have the capacity to integrate such large numbers of rural migrants into the formal economy.[5] As a result, most rural–urban migrants are dependent on largely precarious informal or semi-formal economic activities.

Although the absolute number of people living below the poverty line is still greater in peripheral rural areas, starting in the 1970s poverty became an increasingly urban problem in Africa, not least due to the structural adjustment programs of the IMF and World Bank. As urban living conditions (income opportunities, infrastructure, housing, etc.) have deteriorated in many places, so too have the absorptive capacity of urban centers in a great portion of Africa continually declined.

The dynamic of urbanization in Africa is a process of urban growth without corresponding economic growth; it can also be described as the "urbanization of poverty". This problem finds its spatial expression in the emergence, expansion, and densification of informal settlements, which are now home to the majority of the urban African population. South of the Sahara, almost three-quarters of the urban population now live in slums or slum-like areas (Beguy et al. 2010: 550).

"Urbanization of poverty."
As long as the development of urban employment opportunities fails to keep pace with population growth, the urban lives of many rural–urban migrants will be indelibly marred by insecurity and risk.

Many rural–urban migrants are caught in a kind of urban poverty trap, which they cannot escape due to the precarious labor market conditions and the sharply rising prices for urban housing and food. As long as the development of urban employment opportunities fails to keep pace with population development, the urban lives of many rural–urban migrants will be indelibly marred by insecurity and risk.[6]

---

[5]This process is sometimes described in literature as "overurbanisation" (see Gugler 1982 and Obeng-Odoom 2010).

[6]"Because urban expansion is often not in line with employment opportunities, the urban livelihoods of many people become increasingly insecure" (Cottyn and Schapendonk 2013: 11).

Given the—in some cases extreme—existential uncertainty in African cities, rural–urban migrations very rarely entail the relocation of entire households. Usually, only individual family members move to the cities in search of work. Maintaining close social and economic ties with the regions of origin has been an important aspect of vulnerable urban groups' safety-oriented action since urbanization began in Africa (Potts 2009).

> In this context, household linkages to rural areas have always been an important safety net for many city inhabitants. (Cottyn and Scharpendonk 2013: 11)

Urban sedentism is not the norm in Africa.

The conditions of economic poverty and vulnerability which have persisted for decades in African cities have caused rural–urban migration and urbanization processes to take an altogether different path from that in Europe during industrialization, for example: After a certain period of transience, European rural–urban migration resulted in the permanent settlement of newcomers in urban regions. Accordingly, circular migrations and strong rural–urban linkages marked a transitional phase in the European urbanization process. In Africa, in contrast, urban sedentariness has not become the norm: In most parts of Africa, circular mobility, close contact with the village of origin and the translocal organization of households endure. As Etzold puts it: "Rural-urban migration is often not a singular and unidirectional movement, but rather the first step towards a translocal life" (Etzold 2017: 53). A high proportion of rural–urban migrants maintain strong social, economic, cultural, and emotional ties with their rural areas of origin, often over generations.

> In many African cities there are large numbers of people who have lived in the metropolis all their life, even in the second or third generation, but when asked to name their home, they will invariably give the name of a remote village in the hinterland." (Schmidt-Kallert 2009: 320)[7]

[7]For more on this see Sward's study on Ghana: "Social networks—for the most part—remained active amongst migrants (…), with even second-generation migrants sending remittances, material goods and regularly visiting kin relations in Upper West, showing the apparent durability of such networks." (Sward 2016: 20)

This means that, contrary to the classical models of "*urban transition*" and "*mobility transition*" (Zelinsky 1971), circular forms of migration and intensive rural–urban interconnections are not temporary but rather enduring phenomena in Africa. Urbanization in Africa can therefore only be adequately described in terms of translocality and its strong structuring effect. Politics, planning, and development cooperation must recognize the fact that urbanization and translocalization are parallel processes in Africa and that African urbanity cannot be conceived without translocality.

Urbanization and translocalization are parallel processes in Africa, and African urbanity cannot be conceived without translocality.

5.2 An Attempt at Quantitative Estimation

Despite individual regional differences, the general patterns of urbanization discussed in Sect. 5.1 are applicable to the whole of Africa, making it reasonable to assume that translocal livelihood systems are very widespread.

Quantifying how many people in Sub-Saharan Africa organize their livelihoods in translocal contexts is, however, extremely difficult. No aggregate data on translocal livelihoods are available. And the existing census data, which is deficient in many African countries anyway, gives no indication of the extent or trend of translocal interconnectedness. Another main difficulty is that the census data collected is generally based on the assumption of sedentariness and the definition of households by co-residence (see Douglass 1998: 9). Thus, conventional survey instruments are unable to methodically record translocal household structures. Moreover, such instruments primarily only register international migrations—and very incompletely at that.[8] Resident registration systems, from which reliable relocation data could be derived, do not exist.

Translocal household structures cannot be grasped with conventional census instruments.

> But although the phenomenon cannot be denied, up to the present day official statistics in most countries completely fail to capture part time and seasonal occupations as well as multilocational residence. Temporary or circular migrants do not show up in annual household registration data and are

[8]On the general insufficiency of available data in migration research, see IOM (2013: 35–36, 61–64). For the particular difficulty with the data on internal migration in Sub-Saharan Africa, see Potts (2015).

> even less likely to be counted in a census. (Schmidt-Kallert 2009: 321)

Reliably aggregated data on internal migration processes and their circularity in Sub-Saharan Africa are therefore scarce.[9] Consequently, a quantitative assessment must rely on the findings of individual local and regional studies. Here again, we must bear in mind that translocality is still a comparatively young research perspective (see Chap. 3), which is why there are so few studies of Sub-Saharan Africa available to date that actually refer to the observed phenomena of mobility and space-spanning networks in terms of multi- or translocality. This does not mean, however, that translocal livelihood systems are a new phenomenon in Sub-Saharan Africa. In various studies, some of them older, we find numerous indications of translocal mobilities and network structures, although they are often expressed in different terminology, for example, in studies on *circular migration* (e.g., Bigsten 1988; Elkan 1967; Rain 1999; Potts 2010; Cordell 1996; Dorlöchter-Sulser 2014; Beauchemin and Bocquier 2004; Grawert 1998), *rural–urban linkages* (e.g., Vorlaufer 1992, Potts and Mutambirwa 1990; Forsberg 1999; Baker and Aina 1995; Chant 1998; Tacoli 1998; Muzvidziwa 1997; Ross and Weisner 1977; Greiner and Sakdapolrak 2012), *labor migration* (e.g., Rain 1999; Todaro 1969; Hahn 2004), *remittances* (e.g., Johnson and Whitelaw 1974; Hoddinott 1994; Deshingkar et al. 2015), and *income diversification* (e.g., Reardon 1997; Giesbert 2007). These studies emphasize different issues, but all refer to aspects of translocal livelihoods. Thus, their respective findings can be viewed as traces of the translocal.

"Blank spots on the research map".

Looking at where research on these aspects has taken place, a very uneven distribution of the investigations becomes apparent. There are various studies on Eastern Africa (with a clear focus on Kenya, see Ramisch 2015; Greiner and Sakdapolrak 2012). South Africa is also well researched in this respect (see e.g., Posel 2003; Lohnert and Steinbrink 2005; Kok and Collinson 2006; Ardington et al. 2009; Clark et al. 2007; Collinson et al. 2007; Steinbrink 2009). Other regional focal points include Ethiopia in East Africa (including Devereux 2006; Gibson and Gurmu 2012)

[9]One exception is the Ghanaian *Living Standards Survey*, according to which over 50% of the population are classified as internal migrants—almost half of them women (Awumbila 2015: 133).

and Niger, Nigeria and Burkina Faso in West Africa (Hahn 2004; Youngstedt 2013; Trager 1998; Rain 1999). However, relevant studies regarding the West African countries of Liberia, Guinea or Sierra Leone or Central African countries like Chad or the Central African Republic are few to none.

It would be too hasty to take this as an indication that translocal livelihood systems are only of great importance in those countries which are well studied. Rather, the large blank spots on the African research map—we are convinced—stem not from the absence of translocality phenomena but from the absence of translocality researchers. The fact that research activity differs so greatly from region to region is most likely due to the political and economic situations in the individual countries as well as the academic and personal preferences of the respective researchers.

It is scarcely possible to seriously assess the degree to which the whole African population is embedded in translocal contexts.

From the above, it becomes clear that a serious assessment of the degree to which the entire African population is embedded in translocal contexts is hardly possible. Nevertheless, time and again "translocal numbers" pop up in various studies which can be used as proxy indicators. The compilation of quantitative data from different regional contexts in Table 5.1 is meant to give an impression of magnitude and to serve as the basis for a very rough quantitative assessment of the phenomenon of translocal livelihoods in Sub-Saharan Africa.

While the figures in Table 5.2 reveal some regional differences in the quantitative significance of translocal livelihoods, above all, they offer impressive evidence of how widespread this phenomenon is across the board. These quantitative findings will now serve as the basis for a very rough estimation: Assuming that the results of the individual studies are transferable to the whole of Sub-Saharan Africa and that locally and translocally organized households do not differ significantly in terms of their numbers of members, this would mean that, conservatively estimated, 40–60% of people from rural regions in Sub-Saharan Africa live in translocally organized households—either as "movers" or "stayers". Based on a population of 973 million (World Bank 2016) and a degree of urbanization of approximately 37% (Rauch et al. 2016: 12), this corresponds to a total of 245–365 million people.

The case studies listed above support the assumption that the percentage of the urban population with translocal connections is probably even higher than in rural areas. Working with a rough estimate of 70% of urban dwellers would indicate that another circa 250 million people are embedded in translocal structures.

Table 5.1 "Translocal numbers"—quantitative indicators from individual case studies

	Source	Quantitative indicators
Africa	Tacoli (2002: V)	50–80% of rural households (of all wealth levels) have at least one member in migration
Ethiopia	Devereux (2006: 29)	671 members of the 1100 households investigated in the Somali region qualify as temporary migrants
Botswana	Krüger (1998: 128)	Circa 50% of low-income urban households engage in pastoral or agricultural activities in their villages of origin
Ghana	Awumbila (2015: 133)	>50% of the population can be classified as internal migrants—nearly half of them women
	Ziegelmayer and Spaan (2019: 41)	65% of households indicate out-migration of some members (53% male, 47% female)
Cameroon	BUCREP (2011: 20)	31.2% of the population of Cameroon are internal migrants
Kenya	Agesa (2004: 162)	>33% of Kenyan households are "split" between rural and urban areas
	Mberu et al. (2013)	80% of older residents in Nairobi's slums have strong social links to rural household locations
	Ramisch (2015)	94% of rural households in the study area Ebusiloli (Vihiga County, Western Kenya) have at least one member in migration
	Ramisch (2014: 24)	57% of migrants in Nairobi return at least once a month to their village of origin
	Vorlaufer (1992)	About 80% of Mombasa's rural–urban migrants live in "split households" (between urban and rural)
	Beguy et al. (2010: 553)	82% of the inhabitants of the slum Viwandani (Nairobi) visited their village of origin during the year of the study
	Beguy et al. (2010: 553)	51% of the inhabitants of the slum Korogocho (Nairobi) visited their village of origin in the year of the study
	Tostensen (1986)	88% of migrants surveyed in Nairobi maintained land use rights in rural areas
	Tostensen (1986) (in Ramisch 2014: 15)	54% of migrants surveyed in Nairobi received food from the rural *shamba*
	Paterson (1979: 2)	68% of men between the ages of 18 and 59 resided outside their home area (Bunyore) during the study period
Mali	Tacoli (2002: 19)	70–80% of households in northern Mali have at least one member in migration
	IIED (2001a: 12)	82.5% of the families in Dialakorodji have at least one member who works in Bamako
	IIED (2001b: 12)	71.4% of families in Mopti have at least one member who is a labor migrant

(continued)

Table 5.1 (continued)

Niger	Youngstedt (2013: 142)	82% of Hausa migrants in Niamey have at least one family member or friend in the foreign country
	Dorlöchter-Sulser (2014: 303)	75% of households are dependent on circular migration
	Rain (1999: 151)	95% of the migrants in the city of Maradi have a strong exchange relationship with their home villages
	Gregoire (1992)	70%–80% of urban households in Maradi provide for themselves with agriculture in their home village
Rwanda	Cottyn and Scharpendonk (2013: 14)	90% of the population cultivate at least one piece of land, although only 80% of the population live in the countryside
	Smit (2012: 93)	74.4% of rural households in the Kigali region of Rwanda receive remittances from urban household members
	ACCRON (2011) in Smit (2012: 40)	Approximately 57% of the inhabitants of Kigali are rural–urban migrants
South Africa	ACP (2012: 5)	25% of households have at least one member abroad
	ACP (2012: 5)	40% of rural households have at least one member abroad
	Steinbrink (2009)	>80% of the households in the village studied (Eastern Cape Province) stated that at least one member lived in a city at the time of the survey. On average, 30% of the household members lived in a city
	Lohnert (2002: 240)	An estimated 60% of the inhabitants of former townships and informal settlements in Cape Town live in multilocal household structures
	Clark et al. (2007: 3)	60% of men aged 20–60 in the Bushbuckridge District (Limpopo Province) are circular migrants
	Clark et al. (2007: 3)	20% of women aged 20–60 in the Bushbuckridge District (Limpopo Province) are circular migrants
	Lohnert (2002: 234)	90% of the inhabitants of the informal settlements investigated in Cape Town stated that they travel to their region of origin at least once a year
	Collinson et al. (2003: 8)	55% of households in the Agincourt Subdistrict (Bushbuckridge District) had at least one member in temporary migration
Tanzania	Tacoli (2002: 19)	50–60% of rural households in southern and northern Tanzania have at least one member in migration
	Lerise et al. (2001: n. p.)	>60% of households in the Lindi district, of all levels of wealth, have at least one member in migration

In sub-Saharan Africa, more than half a billion people—more than 50% of the total population—are likely integrated in translocal livelihood systems.

In sum, this means that in Africa south of the Sahara, more than half a billion people—more than 50% of the total population—are likely integrated in translocal livelihood systems.

If this estimate, based on extremely rough values, even remotely approximates the actual order of magnitude, it again becomes clear that politics and development policy and cooperation should be paying much greater attention to the phenomenon of translocal livelihoods in Sub-Saharan Africa.

5.3 Translocal Patterns

In this chapter, the aim is to outline some fundamental patterns of translocal livelihood organization in Sub-Saharan Africa. First, we will explore some spatial and temporal patterns, before discussing some possible regularities in the mobility behavior of different groups and the different motives for migration.

5.3.1 Spatial Patterns

Translocality is ubiquitous in Africa.

Looking at the figures from the case studies available to date (see Table 5.1, Sect. 5.2), translocal livelihoods are clearly very prevalent throughout the whole of Sub-Saharan Africa. An extraordinarily large percentage of the population in cities and rural areas is embedded in translocal contexts. Translocality in Africa is, thus, ubiquitous. Whether the intensity of translocal interconnectedness and the importance of translocal livelihood systems differ from region to region cannot be determined from the present state of research. These questions require further and decidedly comparative studies.

Still, we can identify a few patterns in the geographical orientation of the translocal network structures along which the exchange and mobility of people, goods, information, and capital take place.

Internal migration in African is quantitatively more significant than international migration by far.

With regard to migration processes in Africa, it should first be noted that—contrary to the common media image of the "flood" of African migrants coming to Europe—only a relatively small proportion of the migrations are intercontinental; the large part is intra-African, i.e., international migration between African states and internal migration within them.[10] For example, between 65 and 80% of international migrants from West Africa over the age of 15 have moved within West Africa (Ouédraogo 2009: 129 and Ouédraogo 2012: 1). In Southern Africa, the Republic of South Africa is the most common destination country, hosting about 67% of the region's migrants (Ogude et al. 2018: 2). For Sub-Saharan Africa as a whole, various studies report that two-thirds of international migrants move to other Sub-Saharan states (see De Haas 2010; Fischer and Vollmer 2009; Cottyn and Scharpendonk 2013, UN 2018: 40). Such migrants tend to move within their own regions: "4 out of 5 international migrants in Eastern, Middle, and Western Africa were from the same African region" (UN 2018: 40); in 2017, for example, there was only little migration between East and West Africa. The largest intra-African migration corridors run from Burkina Faso to Côte d'Ivoire, from South Sudan to Uganda and from Mozambique to South Africa (Gandhi 2018). However, internal migration is certainly of greater quantitative significance than such international movements, despite there being hardly any reliable aggregated data available on the subject.[11]

As previously mentioned, the lack of statistical records on domestic migration makes precisely localizing these migrants' regions of origin and arrival extremely difficult. Mobility and the "culture of migration" (see Hahn 2004), especially in the form of circular or temporary migration, are often so ingrained in the organization of everyday life that they largely go unnoticed (by both the scientific community and the public administration):

> Migration is usually considered a last resort, but in fact people often move into town when no work is available in the village and return afterward. The routine nature of this mobility activity is what renders it invisible. (Rain 1999: 7)

---

[10]The UN estimated a total of 214.2 million international migrants in 2011, 18 million of whom came from Sub-Saharan Africa (Cottyn and Scharpendonk 2013: 19). "First, in 1990, African migration was predominantly intracontinental, although the distribution has changed over time. The stock of international migrants originating from Africa and living in Africa was the main stock in 2000, at 12.5 million, highlighting that migration was foremost an intra-African phenomenon. This stock grew consistently, and reached 19.4 million in 2017." (UN 2018: 42).

[11]"Internal migration, between and within urban and rural areas, continues to account for most migratory movements in West Africa." (Black et al. 2006: 23)

Still, we can identify some general trends in the orientation of internal migration systems: Migration mostly takes place from rural areas to urban centers and often from inland regions to the coast,[12] where the larger cities are found in many cases (Ouédraogo 2009: 130).

Translocal livelihood systems: rural–urban as well as rural–rural.

Although the translocal livelihoods organized between rural and urban areas are certainly predominant and, above all, best studied, this is by no means the only constellation. There are numerous compelling indications of strong migration flows and translocal linkages between rural areas, some of which even suggest that rural–rural migration may in some cases play a quantitatively more significant role. Among the studies that signal this is Beauchemin (2011)[13] and Deshingkar (2012)[14] for Burkina Faso; Adepoju (2006)[15] for Lesotho; Mberu (2005)[16] for Nigeria; Greiner (2008) for Namibia.

The basis and motive here are usually translocal economic diversification within the agricultural sector, which often encompasses a combination of small-scale farming (in the area of origin) and wage labor on (large) commercial farms (in the arrival area). In the Somali Regional State in Ethiopia, for example, most migrants who remain within Ethiopia do not travel towards urban centers, like Jigjiga or Addis Ababa (22%), but migrate to other rural areas in the Somali region (62%) (Devereux 2006: 29). Hahn (2004) describes another example: the circular migration from rural regions of Burkina

---

[12]Rain (1999: 114) shows for Tahoua (Niger) that up to 80% of men aged 16–50 move as temporary migrants, mainly towards the coast.

[13]"Among those who immigrated to Côte d'Ivoire, 85% originated from rural areas and only 0.3% migrated within Burkina Faso before leaving their country. These figures confirm the common assumption that Burkinabè migrants usually move directly from rural areas in Burkina Faso to rural areas in Côte d'Ivoire, without other previous movements within their home country." (Beauchemin 2011: 53)

[14]"The Lobi in the south migrate from one rural area to another to find land to rent." (Deshingkar 2012: 3)

[15]"In most parts of Africa, migrants often originate from rural areas and settle and work in rural areas at destination, in part reflecting the predominant feature of rural Africa, the skills profile and occupational structure of these migrants. Cross-border migrants of rural origin also normally work in rural areas at destination, as is the case of the seasonal workers from rural Lesotho who work across the border on asparagus farms in South Africa". (Adepoju 2006: 29)

[16]Based on a national survey, Mberu (2005) shows that the majority (64%) of migrations proceed from rural areas of origin to rural arrival areas (see also Cottyn and Scharpendonk 2013: 23)

Faso towards the plantations in Côte d'Ivoire and Ghana. Currently, Vera Tolo (University of Bonn) is investigating similar forms of rural–rural translocal linkages in Kenya.[17]

In extreme climatic or agricultural situations, there is a stronger tendency towards translocally diversified household management.

The spatial patterns described above are closely related to the motives for migration. In a study on circular labor migration in the West African Sahel zone, Rain (1999) highlights the correlation between climatic and agricultural conditions and temporary migration. According to this comparative study, the most arid regions have the highest migration rates: In the driest area investigated (in the north of the Department Boboye), 28% of adult men are not home at all during the dry season (Rain 1999: 117).[18] In this respect, it can be assumed that there is a stronger tendency towards translocal economic diversification under extreme climatic or agricultural conditions than in more fertile areas. For Benin, Doevenspeck (2005, 2011) shows that processes of soil degradation can also be decisive in increasing rural–rural migration.

The common notion of the "rural exodus" into urban areas must therefore give way to a close examination of the manifold spatial mobilities and networks between various regions of origin and arrival in different contexts.

"step-wise migration".

Moreover, in Sub-Saharan Africa, one also finds patterns of "*step-wise migration*" in which people from rural communities often migrate first to smaller urban centers nearby and then to larger and more distant cities (see Toure 1998; Dorlöchter-Sulser 2014; Doevenspeck 2005, 2011).[19] This pattern can be observed in Burkina Faso, Côte d'Ivoire, Gabon, Mali and Senegal, for example, and corresponds to the classical "*Laws of Migration*" perceived by Ravenstein (1885) (Awumbila et al. 2014). This *step-wise migration* is partly attributable to translocal network developments. When it entails the migration of expanders (see Sect. 4.2.2), *step-*

[17]For details, see: http://transre.uni-bonn.de/en/overview/our-team/vera-tolo/.

[18]Consequently, the income generated through migration accounts for 20% of the rural household income in the northern (arid) study areas, whereas in the southern (and more humid) areas it amounts to only 2–6% (Rain 1999: 117).

[19]"Often, the result is a step-wise migration from rural areas, first to the cities and then abroad. In Côte d'Ivoire, for instance, 28% of the population in 1995 were immigrants, mainly from neighboring Burkina Faso (49%) and Mali (20%) (1993 figures), and most emigrants (78%) moved to West African countries while the remaining 22% emigrated to Europe an America." (Adepoju 2006: 29)

wise migration also marks a spatial expansion of the translocal field, which opens new possibilities for subsequent migrants.

Improving transportation and communication capabilities reshape translocal livelihood systems

While presently no longitudinal studies are available indicating whether translocal livelihoods have tended to span ever greater distances in recent years, this assumption seems very plausible given the developments in transportation systems and, above all, the immense technological advances in telecommunications and new media (including mobile cash transfers) (see Ramisch 2014, 2015; Sterly 2015; Steel et al. 2017). These technological innovations facilitate contact, mobility and translocal transfers, even over long distances.[20] The following passages support this supposition:

> While the African population has been on the move for ages, we can state that African migration patterns become more complex with an increasing diversification in destination and actors. Although a few among many, two factors in particular influence these mobility patterns: first, the decrease in the real cost of travel by the development of transport infrastructure that brings distant places closer and second, the spread of new information technologies which expand the horizons of potential migrants. (Cottyn and Scharpendonk 2013: 22–23)

> Another noticeable trend is the increase in movement to distant destinations, often across national borders. In southeastern Nigeria, destinations include local urban centres such as Aba and Port Harcourt, but also Lagos, and Cotonou in neighbouring Benin; and migrants from Mali can move as far away as the main cities on the Atlantic coast of West Africa, as well as Libya and Saudi Arabia. More affordable transport, increasingly extensive migrant networks and demand in destination countries are some of the main factors for this expansion in the scope of movement. More distant destinations mean that migrants are away for longer periods of time (in Mali, this is often between 5 and 20 years) and, crucially, are unable to return home for the farming season, as is the case with intraregional movement. In northern Mali, labour shortages are increasingly acute and, at the same time, households' dependence on remittances is growing. (Bah et al. 2003: 20)

Another pattern is that the network relationships emanating from the respective villages, which are established in the course of livelihood translocalization, are usually focused on relatively few sites. The translocal networks normally form

[20]See also Tacoli (2002).

clear spatial foci (see Sporton et al. 1999; Doevenspeck 2011), which result from cumulative processes of network concentration through chain migrations ("cumulative causation of translocalization"; see Sect. 4.2.3).

The persistence and flexibility of translocal networks

Once established, social network connections between places exhibit a certain tenacity, and the network configurations are often surprisingly durable. Translocal networks can certainly "move" due to changes in opportunity structures, aligning themselves towards new localities, but this happens quite slowly. Conversely, the translocal networks prove to be very flexible internally. As far as the mobility within established structures is concerned, translocal livelihoods are adaptive and highly responsive: The movements and directions of migration within translocal network structures often adjust directly in response to changing local contexts and opportunities in the various interlinked localities within the translocal social space (climatic conditions, fluctuations in food and rent prices, wage-earning opportunities, political situations, etc.) (see Potts 2010; Steinbrink 2009; Ratha et al. 2011; Schutten 2012). For Zambia and Zimbabwe, Potts (2015) shows that the intensity and direction of internal migration are causally linked to changes in the political and economic environment. In both countries, the dominant direction of migration has reversed at certain times, becoming an urban-to-rural migration flow. People have migrated between agricultural and nonagricultural sectors because of income disparities and moved back to the countryside from the cities in order to reduce the cost of living in times of crisis. This sort of flexible adaptation is only possible because of robust rural–urban social networks, i.e., only because people are organizing their livelihoods translocally.

For that reason, migration movements in Sub-Saharan Africa can by no means be reduced to unidirectional, one-off migration from the countryside to the city; rural–urban migration is not a linear process. Instead, it must be regarded in its full complexity and in light of many factors against the backdrop of the organization of people's lives and household management.

5.3.2 Temporal Patterns

In translocal networks, people adapt their mobility and migration behavior very quickly in reaction to changes in the action contexts at the various interlinked places. These changes are dependent on a variety of political and economic factors at different spatial scales, and thus their occurrence cannot be predicted. Nevertheless, a few temporal regularities can be observed, mainly relating to patterns of mobility over the course of the year. One temporal pattern refers to the (1) seasonality of circular labor migration, another to regularities in (2) visits to the home village in connection with festivals and vacation periods.

Seasonality of circular labor migration

Numerous studies attest that circular labor migration from small-scale farming regions in Sub-Saharan Africa often follows the annual cycle of agriculture, i.e., it is closely linked to sowing and harvesting or the rainy and dry seasons.

> [...] seasonal migration has always been a fundamental feature of Sahelian economy and society and is 'perhaps the principal income diversification strategy, and the most lucrative' (Taylor-Powell 1992: 17). (Rain 1999: 29)

Many able-bodied men and women leave their home villages at the end of the rainy season (e.g., in September in landlocked West African countries) either to take up work in the cities or to "follow the rain" and seek paid employment in another rural area (Awumbila et al. 2014). This form of labor migration often serves the safety-oriented purpose of closing the gap in provision during the agriculturally inactive phases of the year. According to estimates, agricultural activity often only provides enough for about eight months of the year; meaning that the gap for the remaining months must be bridged elsewhere (Oltmer 2015: 13). Seasonal migrants usually return at the beginning of the next rainy season or in agriculturally active phases, as their workforce is again in demand on the farm. Examples of this can be found in Burkina Faso (Oltmer 2015), Niger (Rain 1999), Kenya (Ramisch 2015) and Zambia (Hazell and Hojjati 1995). A similar seasonal pattern can also be seen in child migration, which often follows the patterns and paths of adult migration. Seasonal migration towards the plantations in Côte d'Ivoire and Ghana, for example, affects around 5–6% of children in southern Burkina Faso (Oltmer 2015: 16).

We can also glean the significance of circular migration for the Hausa people, for instance, from their language: According to Rain (1999) there are "25 terms for spatial movement and all but one refer to circular migration of varying duration and purposes" (quoted from Newland et al. 2009: 16).

The seasonal patterns of labor circularity are not only dependent on the agricultural production cycle in the areas of origin though. When it comes to agricultural jobseeking in rural–rural networks, the seasonality in the need for agricultural labor in the target area is also decisive. At the end of their seasonal contracts, the workers return to their villages of origin—regardless of whether agricultural workforce is needed there or not.

Regular visits to the villages of origin

In addition to seasonal circular migration to generate income, most migrants also return to their villages for the purpose of visiting home. Here too, temporal patterns become apparent as the frequency of visits is usually cyclically structured over the course of the year, according to fixed vacation times and holidays. In Christian-influenced regions of Africa, for example, the Christmas and Easter holidays are particularly important "fixed points" in the year. The special significance of the visits at these times is pointed out in studies from Kenya (Ramisch 2015; Beguy et al. 2010), Namibia (Greiner 2010), South Africa (Steinbrink 2009), and Mozambique (Cremin et al. 2015). The volume of road traffic is correspondingly high during these periods when people make their way home.[21] Steinbrink (2009), for example, describes a township in Cape Town, where the resident population in December and January was reduced in some cases to about one-third, while the number of those staying in the rural research area (a village in the Eastern Cape Province) almost doubled at the same time.[22] This marked seasonal shift in the whereabouts of large sections of the population can be observed in various African countries. Cremin et al. (2015) report one notable *indirect* effect of this temporal pattern of translocality: In the Gaza Province in Mozambique, for example, there is a significant increase in the birth rate in September of each year (nine month after the Christmas holidays), a trend that is not discernible in parts of the country with lower rates of labor migration (Cremin et al. 2015: 2).

[21]In many regions of Africa, the number of deaths on the roads rises sharply during holiday periods. This is due not only to the larger number of travelers but also to serious shortcomings in safety conditions in the transportation sector and holiday-induced alcohol abuse.

[22]A resident of Township Site 5 describes the atmosphere in the urban settlement at the Christmas season as follows: "You know, in December Site 5 is a little bit like a ghost town, like a graveyard … It's quiet, man, many doors are locked. All the people are at home, you see, in Transkei." (Steinbrink 2009: 279)

Functions of festivities in the home village.

Funerals also plays an extremely significant role in translocal social structures in Africa. In many regional contexts where the veneration of the dead is of great importance, it is a sacred duty to bury the deceased in the "soil of the ancestors" (regardless of where they died) and to honor both the deceased and the ancestors at the funeral. This tradition contributes substantially to the stabilization of translocal ties.

Often, other (partly regional and culturally specific) celebrations and rituals such as weddings, baptisms and circumcisions are held during the annual holiday and vacation seasons too (as observed, for example, in Cameroon: Pursey 2014; Namibia: Greiner 2008, 2010; Niger: Rain 1999 and Burkina Faso: Piot 1999). Rituals that take place at greater intervals can also be important for translocal cohesion.[23]

To return for such events is widely regarded as a moral imperative and/or social obligation, which migrants are undoubtedly obliged to fulfill.[24] Migrants often bring gifts of money or goods to the family members in the countryside on these occasions. But even "hapless" jobseekers return to the villages at such times—often hoping to receive information from the "more successful" migrants about job opportunities in other places (see e.g., Steinbrink 2009). In translocal networks, the place of origin usually forms the central spatial node, and every festivity is an important opportunity for face-to-face contact between members of the translocal community, who frequently reside very far from each other otherwise.

Holidays, festivals and rituals in the migrants' area of origin therefore have an immense significance for translocal cohesion, because they represent reliable opportunities for

[23]De Jong (1999) provides an example of this for Senegal: The initiation ritual "*bukut*" is fundamental to the cohesion between migrants and those who remain at home, although it only occurs every few decades. All men of the village community born since the last *bukut* are meant to take part in this ritual. All migrants return to their home village for this event, even those born elsewhere. The whole village community contributes to the *bukut*: Those earning money remotely send this or other material goods home in order to compensate for the loss of agricultural activity in the year of the *bukut*. De Jong (1999) describes how this ritual of initiation reinforces the relevance of the home village among migrants, strengthening translocal social bonds.

[24]"You are supposed to go back home at least after 1 year, perhaps in the big holiday between December and January. You are supposed to go back and meet the whole family." (Interview with a rural–urban migrant in Cape Town, Steinbrink 2009: 397)

strengthening existing social ties, build new ones and exchange information.

In addition to the seasonal patterns, there are certain temporal patterns within the translocalization process itself, like the diversification of the motives and types of migration over the course of a translocal network's "maturation", the ages at the time of migration and remigration, etc. These aspects will be discussed in more detail in the following chapter.

5.3.3 A Phase Model of Translocalization: "From Expanders to Transmigrants"

The forms and motives of migration and the groups of people involved are subject to change in the course of the translocalization process.

Translocal livelihoods are not simply the result of migration; in fact, translocality itself induces migration. In the context of translocalization, the forms and motives of migration and the groups of people involved shift over time. Considering these changes can help us to better understand migration in Africa.

For that reason, we present here an ideal-typical three-phase model of the emergence, consolidation and maturation of translocal networks (see Fig. 5.1). The model illustrates which forms of migration and which "types of migrants" predominate in different phases of translocalization.

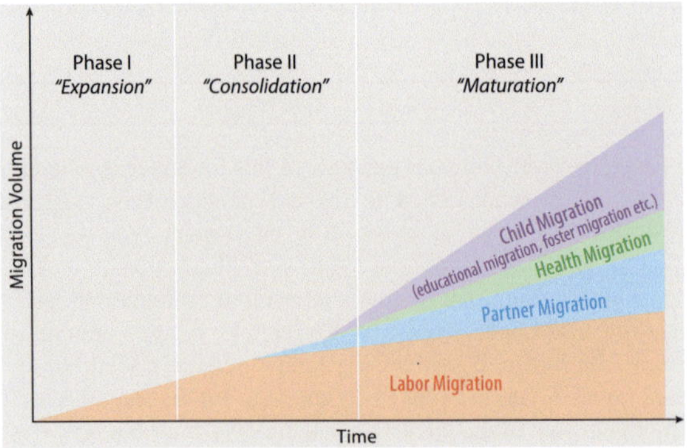

Fig. 5.1 Phase model of translocalization: diversification of migration over a translocalization process. *Source* own illustration

Phase 1: Translocal expansion by "typical labor migrants"

Phase 1: "Translocal Expansion"

The phase of translocal expansion describes the beginning of a translocalization process between two sites. Here, pioneer migrants or socio-spatial expanders (see Sect. 4.2.1) play a pivotal role. These migrants do not cut their social ties to relatives and friends in the region of origin, instead of maintaining these through remittances, visits, the exchange of information, etc. Many studies support the assumption that these expanders are generally "typical labor migrants": The search for work is the main motive for their migration. This phase primarily entails adjusting the economic activities of households for the purpose of economic diversification with the aim of crisis management or risk minimization.

A review of the research literature suggests that most expanders are young men: "*Usually, only a few members of rural households, mostly men, move initially to the cities*" (Krüger 1998: 121). The socio-spatial expanders form the crystallization nuclei of translocal network building; they prepare the translocal field through their migration.

Phase 2: "Consolidation" (Transmigrations Beyond the Household)

Phase 2: Network consolidation

In the second phase, the phase of consolidation, a self-reinforcing process of chain migration is slowly set in motion. The expanders (see above) serve as points of contact for further migrants from the region of origin. In phase two, labor migration is still the principal mode. Subsequent job-seekers use the now-established translocal networks to reduce the risks of migration. Their contacts at the target location can facilitate access to housing and assist with the job search, for example.

During this phase, it is primarily friendship and kinship ties beyond of the limits of the migrant household that are utilized, as the translocal networks within the households are not yet sufficiently developed. Young male migrants also predominate in this phase.

Phase 3: "Maturation" (Transmigrations Within the Household)

Phase 3: Maturation of the translocal network:

- *more female migration*
- *more child migration*
- *more migration of elderly and sick*

In the third phase ("*maturation*"), there is an increase in transmigration within the household. Once the first migrant of a translocally organized household has established him- or herself in a certain place—i.e., has attained relatively secure housing and income—other household members often follow, making use of the intra-household network relationships.

In this phase, the forms of migration become more diverse; other motives crop up. While the search for work often remains an important motive, migration diversifies markedly. Over the course of this "*maturation phase*", other forms may even supersede labor migration in their quantitative significance, thus changing the composition of the group of migrants:

More women migrate: Women presumably act very rarely as socio-spatial expanders. The migration of women tends to mainly take place within already established translocal fields and mostly within translocalized household structures. As a rule, their motives for migration are somewhat more diverse than those of men: Women's migration may take the forms of labor migration or "partner migration", or it may involve caring for other household members at the urban household location. The latter is particularly the case when children reside there.

More children migrate: Children are also mainly mobile within established translocal networks and within translocal households. The migration of children in translocal networks is, however, a little researched aspect, although their migration probably accounts for a considerable, but clearly underestimated, proportion of internal migration movements in Africa.

Depending on the specific situation at the different household locations (e.g., in rural and urban areas), many *small children* experiences a change in their place of residence within the first years of life. Many then spend their early childhood years at different places—partly with and partly without their biological parents.

Schoolchildren are often mobile within translocal network structures. Some children and young adolescents are sent to household members or other relatives in urban areas to attend school, but educational migration also takes place from the city to the countryside. Beguy et al. (2010), for example, point out that many residents of slums send their urban-born children to rural areas because the educational conditions where they live are considered too poor. For South Africa, Lohnert (2002) reports that many young couples and unmarried mothers move to urban centers, but in over 50% of cases, their children stay with the grandparents in the rural Eastern Cape Province during the school year and only come to their parents in the city during school holidays.

The translocality of livelihoods likely gives many children in Africa better access to (better) schooling than they would have otherwise received. This is particularly true if there are no (secondary) schools in the regions of origin. The transition from primary to secondary school is often only possible through migration, and this form of educational migration is usually only possible within translocal networks (for more details, see Sect. 6.3.1.).

Child labor migration occurs frequently in Sub-Saharan Africa as well. Some children leave their villages of origin in order to contribute economically to the livelihood security of their households. There are only a few recent studies on this aspect to date. For Ghana, Kwankye et al. (2009) and Tamanja (2014) provide indications of this form of child migration, as does Oltmer (2015) for Burkina Faso.[25]

More elderly and sick people migrate: The migration of elderly or sick people mainly takes place from the city to the countryside. Many older migrants state that, at the end of their working life in the city, they wish to spend their remaining years in their home village (see, among others, Devereux 2006; Greiner 2008). From an economic viewpoint, this form of migration is often also about minimizing the cost of living, which is generally higher in cities than in rural areas. However, health-related migration also flows in the other direction, as older and sick people sometimes go to the cities—usually only temporarily—to receive modern clinical medical care and necessary medications (Steinbrink 2009). This form of migration also primarily takes place within translocal household structures (see Sect. 6.3.3.5).

Moreover, in the third phase, the established (household internal or external) translocal social network structures can be used for sudden migrations in response to crises at the rural household location or in the region of origin. A stable translocal network can thus have a mitigating effect on crises

[25]"Thus, as mentioned before, there often is an agreement between parents and their children on migration decisions; in several cases the parents even prompt their children to migrate and in addition, pay for their transport down south, perhaps with the hope of receiving remittances either in cash and/or in kind from their migrant children. As noted below, it was found that a significant share of the child migrants surveyed do send home either money or other goods, or both. [...] Even though there was no evidence of any informal agreement between child migrants and their parents for the former to remit to the latter, many of the child migrants indicated that they do remit to their parents in cash and/or in kind. The in-kind remittances are two-way, where child migrants receive some maize, local rice, yam, groundnut, and groundnut paste from their relations, especially their parents. They, in turn, send back clothes, pieces of cloth, lace material, shoes, and sandals and items such as cooking utensils, soap, milk, tea bags, Milo, sugar, and imported rice. Some of the children are able to send substantial amounts of money at a time. However, this is not done on a regular basis and it depends on the availability of a trustworthy person going home." (Kwankye et al. 2009: 28)

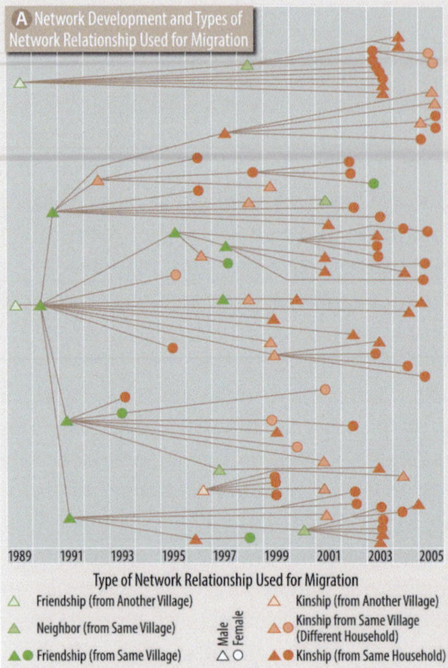

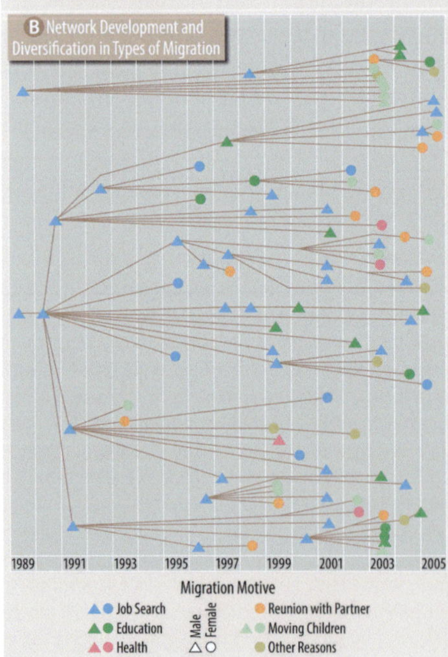

The two network diagrams (A, B) illustrate the development of a migrant network and the diversification in types of migration over different phases in the translocalization process between a village in South Africa's Eastern Cape Province and an informal settlement (Site 5) in Cape Town. The symbols stand for the migrants, who arrived over time (1989–2005); the network lines indicate who the first point of contact was.

The first migrants *("socio-spatial expanders")* served as crystallization points of translocal network development. While at first friendships were mainly used for migration to Site 5, other social relations played a greater role as the process progressed. In the period after 1998, relatives were the primary point of contact. At first, more distant relations within the kinship group were of particular importance; from 2002 on, however, migration within translocal households clearly prevailed (see A).

There are two reasons for this tendency: First, social norms regulate which type of relationship has priority when it comes to receiving certain forms of support. Second, the tendency correlates with the shift in the dominant types of migration over the course of the translocalization process (see B): The first migrants to arrive mostly conformed to the "classic *labour migrant*" profile – young men looking for work (see PHASE I in "Phase Model", Fig. 9). Only in the subsequent phases (PHASE II and III in Fig. 9) did a greater number of women migrate to Site 5 and the motives for migration become more diverse. For example, from 1999 on, schoolchildren were increasingly sent to Site 5 to make use of the better educational opportunities there *(education migration)*. Some wives and fiancées of migrant workers also moved to Site 5 *(partner migration)*, many of them with small children. Other children were placed in the care of relatives at Site 5 *(foster migration)*. Elderly and sick people moved to Site 5, too, to access medical care in Cape Town *(health migration)*. These latter types of migration mainly arise in later phases of network development (Phase II, III) and fundamentally take place within the context of the translocal household.

As these figures illustrate, the utility of the translocal network expands as it matures: a well-established network offers more than just access to the labor market. Though labor migration (with the aim of economic diversification) is the initial trigger of the translocalization process, different forms of migration arise as a result. The social translocalization signifies the development of a translocal network structure that directly affects very diverse aspects of livelihood organization.

Fig. 5.2 Phase progression of translocalization—an empirical example. *Source* own illustration adapted from Steinbrink (2012: 185, 186)

and strengthen the resilience of the actors and livelihood systems involved.

Immigration into urban areas beyond the search for employment (education migration, health migration, migration of young children, etc.) primarily entails the migration of economically dependent individuals, who continue to depend on the economic support of others long-term, even after their migration. It is therefore not surprising that the vast majority of these migrations take place within the narrower family–social context of translocal households. Despite the lack of data, there is a reason to believe that these forms of migration now account for a significant proportion of internal migration in Africa.

Figure 5.2 illustrates the diversification of migration over the phase progression of translocalization, based on an empirical example from South Africa.

5.3.4 *Socioeconomic Patterns: "Who Stays, Who Leaves, Who Returns?"*

As shown above, in the course of the emergence and establishment of a translocal livelihood system, the patterns of who migrates and who remains at home shift over the different phases and reflect different reasons for migrating. In this respect, there are no general types of *movers* or *stayers*; ultimately, the question of *who stays, who leaves and who returns* can only be answered with regard to factors like the "maturity" of the translocal network and the motives for migration.[26] Nevertheless, we will outline a few patterns here.

5.3.4.1 Gender

Research on migration in Africa has largely focused on phenomena of (circular) labor migration. Other forms of migration have long been neglected. Thus, the conviction has prevailed in previous research that migration in Africa is predominantly a "male affair".

---

[26] "[…] the important role played by migrant networks has been clearly documented in work on the environment-migration nexus, determining both who migrates (Apeldoorn1981; Haug 2002), and where they go." (Morrissey 2013: 1502)

Classical variant of male labor migration still predominates in numbers.

Indeed, the classic variant of male labor migration still seems to predominate in terms of numbers. There is plenty of evidence for this.[27] Devereux (2006: 29), for example, reports in a study from the Somali region in Ethiopia that the temporary migration rate among men is significantly higher than among women.[28] In his study on Hausa communities in Niamey (Niger), Youngstedt (2013) also describes men as being more likely to migrate (Youngstedt 2013: 136, 148). According to the author, in 20% of rural households in Niamey, women are the de facto heads of household, meaning that the burden of additional income also falls on them if their husbands' remittances fail to materialize. Greiner and Sakdapolrak (2012) observed a similar pattern in Kenyan Machakos. During the droughts of the 1980s, about 60% of rural households were headed by women, which led to a simultaneous increase in the "feminization of agriculture" (see Sect. 6.3.2) and in domestic responsibilities, as well as an increased dependence on remittances sent by the husbands (Greiner and Sakdapolrak 2012).

Demographic imbalance in the gender ratio in urban and rural regions due to male labor migration

In many parts of Africa, male labor migration has led to grave demographic imbalances in the gender ratios of resident populations, both in the (rural) regions of origin (see Fig. 5.3)[29] and in the (urban) arrival regions.[30]

---

[27]For example, Lerise et al. (2001) describe young unmarried men as the most mobile group in Lindi (Tanzania). Affirmative evidence can also be found in Burkina Faso (Hahn 2004: 384), Nigeria (Okali et al. 2001: 28) and Rwanda (Smit 2012: 76).

[28]"(…) differential mobility between males and females, with young to middle-aged men, in particular, being more likely to travel for extended periods than women." (Devereux 2006: 29)

[29]Oltmer (2015: 13) shows that in some rural regions of Burkina Faso there are just 70 men per 100 women. Regarding the quantitative gender ratio in the Somali Regional State of Ethiopia, Devereux (2006: 29) states that there is a clear female predominance among 20- to 39-year olds in the rural resident population. Between the ages of 40 and 59, on the other hand, there are more men, and there is a still greater imbalance of 200 men to 100 women aged 60 and over (Devereux 2006: 29). This suggests increased out-migration of men between the ages of 20 and 39 and a return from the age of 40 (to retire at home).

[30]"High urban sex ratios, a substantial preponderance of males, are reported from a number of African countries. The extreme examples are Zimbabwe, 1,412 males for 1,000 females in 1969, Kenya, 1,386 males for 1,000 females in the same year, and Namibia, 1,268 males for 1,000 females in 1960—not surprisingly in view of their history as settler colonies." (Gugler und Ludwar-Ene 1995: 259)

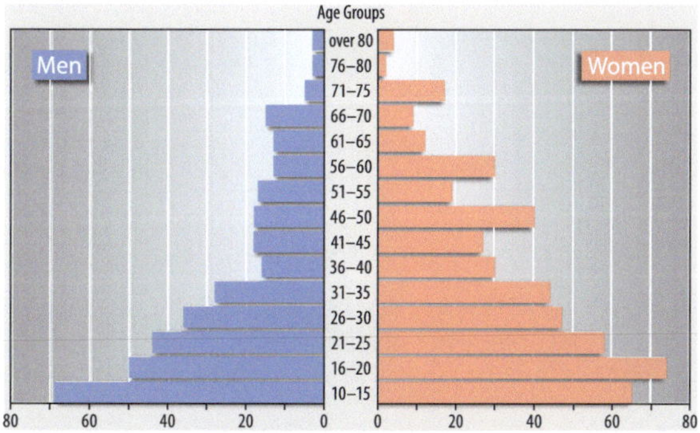

Fig. 5.3 Examples of demographic imbalance in gender ratio in a rural region of Rwanda. *Source* modified from Smit (2012: 76)

> As a result of the great majority of male migrants of working age, the rural area of origin is left with a demographically unbalanced population of women, younger children and older people. (Smit 2012: 76)[31]

"Feminization of Migration".

However, there are many signs that these inequalities are diminishing now and will continue to do so in the future as more and more women migrate.[32] For West Africa, Adepoju (2005: 2) puts it thusly: "Traditional male-dominated short-to-long-distance migratory streams in West Africa are increasingly feminised". Youngstedt (2013: 134) confirms the trend specifically for female Hausa immigrants from Niger: "I recognize that Nigerian Hausa women navigate complex migratory trajectories that illustrate contemporary global trends of the 'feminization of migration.'" And Cottyn and Scharpendonk (2013) write of Ghana[33] and Tanzania[34]:

> While traditional migration regimes in sub-Saharan Africa used to be male dominated, contemporary mobility patterns are becoming increasingly feminised, resulting in ever greater shares of women, particularly younger women, who migrate […]. (Cottyn and Scharpendonk 2013: 3)

[31]The sex ratio (proportion of men in comparison with the proportion of women, multiplied by 100) for the ten to fifteen age group is 105.8, but for ages sixteen to fifty it decreases to 64.4, demonstrating a large surplus of females. (Smit 2012: 76–77)

[32]See also Gugler and Ludwar-Ene (1995).

[33]See Addai (2011) as well as Castaldo et al. (2012).

[34]Cottyn and Scharpendonk (2013: 29) indicate in their research on Tanzania that one in three households now has a migrated female household member.

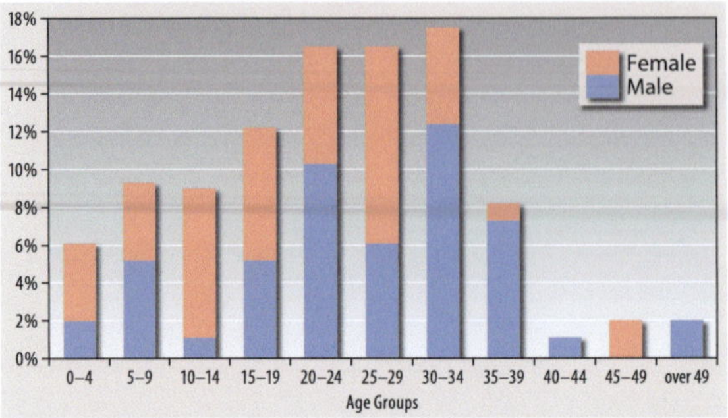

Fig. 5.4 Age distribution of rural–urban migrants (by gender) in an informal settlement in Cape Town. *Source* modified from Steinbrink (2009: 278)

In relation to Southern Africa, Crush et al. (2005: 6, 14–15) also speak of a "*feminization of migration*" (see Sect. 6.3.2).

We argue, that with the consolidation of translocal networks, more and more women are being incorporated into migration processes. Using a bilocal case study, Steinbrink (2009) demonstrates how the proportions of male and female migrants have converged over a period of 17 years in the course of chain migration and the "maturation" of the translocal network between a rural area in the former Transkei, South Africa, and an informal settlement in Cape Town.

There is often a *time lag* between male and female migration in the translocalization process, because women are much less likely to be among the early migrants ("expanders"); they tend to migrate within established networks, even when migrating as jobseekers. With the diversification of the forms of migration (see Sect. 5.3.3), however, female household members become increasingly mobile. The age distribution of a group of rural–urban migrants from one village of origin in an informal settlement in Cape Town (see Fig. 5.4) can be understood in this light. The age structure among the group of migrants shown in Fig. 5.4 is clearly dominated by the economically active age groups. At 54%, however, the share of men is only slightly higher than that of women. It is evident that the proportion of female migrants is higher in the younger age cohorts in particular. So, with a certain delay, more women are also moving into the city.

Further examples of women migrating in the context of translocal household structures can be found in studies on

Kenya (Beguy et al. 2010), Tanzania and Ghana (Castaldo et al. 2012). For Tanzania, Tacoli and Mabala (2010: 391) found that over 30% of the households surveyed had at least one female member living elsewhere at the time of the study.

Probability of women migrating decreases as number of children increases.

Beguy et al. (2010: 571, 576–577) describe a particularity of female migration in translocal contexts in their study of two slums in Nairobi. The authors argue that women have a higher frequency of mobility between the rural place of origin and the city than men.[35] In addition, they show that the reasons for migration among women tend to be more directly family-related (e.g., marriage, child-rearing, caregiving, etc.) (Beguy et al. 2010: 571, 576–577). Brockerhoff and Eu (1993) stress the importance of family events (such as marriage, and childbirth) to women's mobility in Sub-Saharan Africa and point out that women with small children are less likely to move to urban areas. Reed et al. (2010: 793) make the same observation: Married women with more than two children migrate significantly less frequently. In South Africa, Steinbrink (2009) and Clouse et al. (2018) observe that young women who move to urban areas as schoolgirls or migrant workers and become pregnant usually stay there during pregnancy in order to take advantage of the better medical care in the city. Only after childbirth do they move to the rural household location with the newborn child, in order to access the familial support of the female household members there. When the children are a little older, the young mothers often go back to work in the city, leaving their child at the rural household location, often in the care of the grandmother.

In Ghana, too, women migrate primarily for family and marriage reasons (81% in sum) and less often for gainful employment (11%). As a motivation for migration, education is similarly low for both sexes in Ghana (4% for men, 3% for women) (Castaldo et al. 2012: 12). These studies thus confirm the changes in forms of migration over the course of the translocalization process described in Sect. 5.3.3.

---

[35]Devereux (2006) also shows for the Somali region in Ethiopia that the tendency towards longer absences is higher among male migrants than among female migrants. The "Migrating Out of Poverty" survey in nine Sub-Saharan African countries also shows that women migrate for shorter periods than men do (UN 2018: 71).

The migration of women is, however, by no means limited to "directly family-related" mobilities as economic dependents. Rather, a significant proportion of female migration does in fact take the form of work migration. Werthmann (2007) describes an interesting example in her study on the labor migration of women to the gold mining areas of Burkina Faso. These women are often unmarried or divorced mothers. Werthmann (2007: 307) reports that women sometimes leave their children with their parents in order to work in the gold mining areas. In this case, the translocal network does not primarily span between husband and wife but rather between the female migrant worker, her parents and her children (Werthmann 2007: 307). The degree to which the overall feminization of migration can be attributed to a change in gender roles must still be examined (see Sect. 6.3.2).

The assumption that it is primarily men who migrate is by no means fully in line with reality.

All in all, it is evident that the long-prevailing notion in development research and planning that it is primarily men who migrate, while women remain behind in the village, can by no means be categorically affirmed for all regions of Sub-Saharan Africa.

5.3.4.2 Age

Within the context of translocal livelihoods, we also find certain patterns in the spatial distribution of different age groups. These can be explained by differences in migration behavior—in different phases of life on the one hand, and in the different phases of the translocalization process (Sect. 5.3.3) on the other hand.

Probably the most typical and widely researched pattern is the migration of younger people of working age (usually between 15 and 45 years) to urban areas in search of paid work. As a result, it is the very young and the very old who stay behind, whereby those adults of working age who remain at the rural household location also play an important role, not only in small-scale agriculture but also in raising children and caring for the elderly and the sick[36] (Schmidt-Kallert 2009: 326–327; Greiner 2010: 142).

Typical pattern: labor migration of young people of working age

[36]Daily child-rearing and caregiving tasks are still predominantly performed by women.

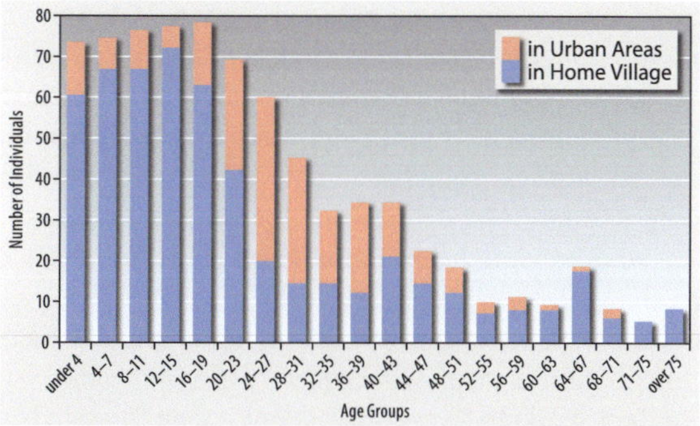

Fig. 5.5 Age distribution of a translocal community (by whereabouts). *Source* modified from Steinbrink (2009: 221)

Many studies address this widespread pattern and the resulting demographic imbalance in the age structure, including, among others, studies of South Africa (Lohnert 2002; Steinbrink 2009), Ethiopia (Devereux 2006), Burkina Faso (Cordell 1996), Niger (Youngstedt 2013, Rain 1999), Namibia (Greiner 2010), Ghana (Ziegelmayer and Spaan 2019: 41) and Kenya (Beguy et al. 2010).

Figure 5.5 shows this "anomaly" in the age structure caused by translocal migration for a village in the former Transkei, South Africa. About 60% of the residential population in the village is of school age or younger; almost 10% are 60 years or older. About half of the middle-aged, i.e., most economically active, population has left the village (temporarily) and lives in urban areas. This proportion is highest in the age group of 24–43 years (73%).

Similarly, in his case study on Hausa migrants in Niamey, Niger, Youngstedt (2013: 137) concludes that most men (and women) between the ages of 15 and 45 are "*on the move*" or "*in transit*". Further, for Tahoua in Niger in the 1980s and 1990s, Rain (1999: 114) describes that up to 80% of men between the ages of 16 and 50 can be regarded as temporary migrants. Devereux (2006) describes a similar situation for the Somali region in Ethiopia, citing that mainly "*young to middle-aged men*" have migrated. Beguy et al. (2010: 563) report this "classical" pattern too in their survey of movements between two slums in Nairobi and either the rural home communities or other urban districts: They show an increase in the migration rate for those aged 14–24 years, followed by a diminishing rate.[37] Likewise, Cordell (1996: 293–294) confirms for Burkina Faso that it is primarily young adults who enter into labor migration and then return home at middle age. He problematizes the emigration of the young and strong from the countryside, which he sees as creating extraordinary difficulties for the very old and very young who are left behind struggling to secure their daily survival.[38]

Older people often migrate back to rural household locations at the end of their working lives.

In rural areas of origin, the demographic imbalance in the age distribution is also a consequence of *remigration*: At the end of their working lives, older people often migrate back to the rural household locations in order to spend their "retirement" there.

"Myth of Return"

Many migrants state that they want to spend their old age at their native home (see, among others, Devereux 2006; Greiner 2008; Steinbrink 2009). However, it should be mentioned that this "desire to return", which has been repeatedly documented in numerous studies and different regional contexts, sometimes reflects social expectations rather than the individual's actual wish to return.[39] The "*myth of return*" certainly exists in a translocal context as well. Steinbrink (2009) points out that the discrepancy between the social desirability of return and the actual wish to return is even more pronounced among female migrants than among men. He explains this disparity through the traditional gender division of labor in rural areas, which makes "country life" significantly less attractive for women.

From an economic perspective, the remigration of the elderly, the sick and the unemployed is often also aimed at minimizing the cost of living, which is generally higher in cities than in rural areas (see Sect. 6.3.3).

The remigration of the elderly and sick is further justified by the cultural necessity of dying or being buried in the village of origin—among the ancestors (see examples for Kenya/Nairobi: Beguy et al. 2010: 564; for Namibia: Greiner 2010: 142; for Niger: Youngstedt 2013: 137–138; for South Africa:

[37]Collinson et al. (2007: 4–5) show similar results.

[38]On this, see also Oltmer (2015: 13–15).

[39]In his study on Ghana, Sward (2016) reports that the desire to return is relatively rare among migrants—apart from older migrants who want to spend their old age in rural areas. These are usually older men who assume the role of the head of the family in the countryside and must thus subordinate their individual interests to the interests of the community (Sward 2016: 24).

Steinbrink 2009). Accordingly, Okali et al. (2001: 24) report from the region around the Nigerian city of Aba that the body of a migrant who dies in the city is taken to the village of origin for burial. This explains why there are funerals almost every weekend in rural areas, which in turn gives the impression that the mortality rate in the countryside is significantly higher than in the city. Again, from an economic point of view, it is much more cost-effective to return to the country during one's lifetime, as transporting a corpse is often many times more expensive than the transport of a patient (see Steinbrink 2009). This also explains higher mortality rates in rural areas compared to urban areas. According to Okali et al. (2001:24), this "tradition of return" makes the rural area of origin "*a place to come and die*" (statement of an interviewee in the study). However, Youngstedt (2013) offers evidence that this form of remigration is losing importance. He observes that today more and more migrants also spend their old age in the city (see Youngstedt 2013: 138).

Child migration as independent labor migration or as a result of the parents' labor migration

With regard to age structures, the *migration of children* is also relevant. According to the three-phase model presented above (see Sect. 5.3.3), children become increasingly mobile within the translocal network at a later point in the translocalization process. As with adults, this translocal child mobility is partly aimed at finding gainful employment for the children themselves (see e.g., Kwankye et al. 2009 and Tamanja 2014), but it can also result from the labor migration of their parents.

"foster children"

In the latter case, children are often entrusted to other household members. Research on "*foster children*" can be found in numerous studies (e.g., Bledsoe et al. 1988 for Sierra Leone; Isiugo-Abanihe 1985 for Sierra Leone, Ghana, Liberia and Nigeria; Greiner 2010 and Klocke-Daffa 2015 for Namibia; Hashim 2007 for Ghana; Pilon 2003 for West Africa; Lachaud et al. 2016 for Burkina Faso; Martin 2015 for Benin). Greiner (2008) finds that 40% of children in the rural peripheral area of Fransfontein in Namibia are "*foster children*"—most of them of preschool age, sent by their urban-dwelling parents to their grandparents in the rural area. When the children enter school, they often migrate back to the city—marking the start of a lifelong "migration biography":

> A frequent change of residence [of children] is common: By the end of their school career, most children have lived in several towns or cities and with different relatives […]. (Greiner 2010: 144)

Lohnert (2002) presents similar findings for South Africa: She states that 50% of the children of rural–urban migrants do not live at the urban household location. Frequently, the city only becomes a relevant place of residence for the children when they begin attending school (Lohnert 2002: 239; see also Sect. 6.3.2).

Although the practice of "*child fostering*" has received attention in many studies, the relevance of this form of migration to maintaining and expanding translocal livelihood networks is hardly ever referenced. Still, we can assume that it is an aspect of considerable influence.[40]

5.3.4.3 Level of Education

Education and school attendance are an important migration-inducing factor in African translocal life contexts.

As made clear above, education and school attendance represent an important migration-inducing factor in translocal life contexts in Africa: People migrate to generate monetary income, which is also used for children's schooling; children are sent to household members or other relatives who live somewhere else to gain access to (better or more advanced) schooling; and adults—mostly women—migrate to the place where the pupils live to care for the children there.

However, the question of whether there is a connection between educational levels and migration or translocality cannot be answered unequivocally based on the available studies. Nevertheless, Beguy et al. (2010) demonstrate with an example from Nairobi that less educated people tend to migrate less frequently to the two slums Viwandani and Korogocho than those who have at least a primary school education. Further, they observe that the less educated rural–urban migrants are also less mobile than those with a higher level of education, and their stay in the slums tends to be longer than those of the more educated. Accordingly, people with (higher) school education have a somewhat higher probability and frequency of migration (Beguy et al. 2010: 570–571).

An older study by Gould (1989: 269–270) on Tiriki in Western Kenya examines the influence of (technical) training

[40]At the *Institute for Migration Research and Intercultural Studies* (IMIS) at the *University of Osnabrück*, Germany, Hannah Niedenführ is currently preparing her doctoral thesis on child migration and the system of "*confiage*" in- and outside of translocal networks in Burkina Faso.

on migration behavior. The author concludes that education functions as a kind of preparation for migration:

> The main balance of opportunity remains outside the area, and education, whether in academic or technical schools, is felt to be a necessary part of the preparation for migration. (Gould 1989: 270)

According to Gould (1989), the educational programs originally implemented to promote rural development resulted in a "culture of migration", which in turn created a great demand for better schooling. Waddington and Sabates-Wheeler (2003) also assume that there is a positive correlation between a higher level of education and the probability of migration, reasoning that better employment opportunities (in the city) motivate the potential *movers*:

> Rationality implies that individuals with better education, skills and labor market experience have a comparative advantage in job search at destination labor markets, and therefore are more likely to migrate. Thus, migration is seen as a selective, rather than random, process, and whilst migrants ›self select‹ in this way, the same logic of rationality implies that nonmigrants do not move because their comparative advantage lies in staying (Tunali 2000). Consequently, it can be derived that other things being equal, migrants are more likely to be from ex ante better-off groups, who are more likely to be better educated and skilled, and nonmigrants from worse-off groups. (Waddington and Sabates-Wheeler 2003: 5)

Gibson and Gurmu (2012) make similar arguments in a study of Ethiopia.

Whether the probability of remigration in old age is lower among more highly educated migrants cannot yet be answered based on existing studies. If, however, one assumes that there is a connection between higher education and vocational success, and if one also assumes that remigration takes place at least partly for economic reasons (minimizing the cost of living) (Sect. 5.3.4.2), then the assumption appears plausible.

Because the reasons for translocalization are manifold and not bound to a universal action logic oriented towards maximizing benefit or minimizing risk (see Sect. 4.2), the connections between education and the translocalization of livelihoods are by no means unequivocal either. There is a clear need for more research on this point.

5.3.4.4 Level of Wealth or Poverty

Both poorer and better-off households are involved in internal migration, though with different motives and "outcomes".

To what extent the level of household prosperity is decisive for the likelihood of livelihood translocalization cannot be incontrovertibly answered with present literature either. The available studies largely deal with the differences in migration probabilities and migration motives between poorer and wealthier households. There is a broad consensus here that both poorer and better-off households are involved in internal migration in Sub-Saharan Africa, but each has different motives, and their "*outcomes*" differ.

Generally, though, there is disagreement as to which wealth group is more likely to migrate. Waddington and Sabates-Wheeler (2003: 5), for example, argue that the members of financially better-off households have a higher probability of migration because they tend to have a higher level of education and thus better chances upon entering the (urban) labor market. Gibson and Gurmu (2012: 5) draw this conclusion too in a study of Ethiopia.

Other authors, however, take the opposite view. Among others, Olson et al. (2004: 22) conclude in their study of Kenya that members of poorer households are more likely to leave their home village.

For his part, Smit (2012: 72) finds no connection whatsoever between the level of prosperity and the probability of migration in his study of the region around Kigali (Rwanda), arguing that the very distinct migration motives among the different wealth groups account for this. Referring to Schutten (2012), he states that 50% of the migrants in the study area were literally forced into migration because of their extreme poverty, whereas 24.6% moved to the city to improve the living conditions of their households and another 25.4% stemmed from the better-off households and migrated to the city to further increase their prosperity.[41]

[41]"There was even no relationship between social-economic household status and the presence of a migrant household member. The explanation for this lies in the fact that migrants come from each social-economic household group and migrate with different reasons. According Schutten (2012) namely 50% of the migrants in the research areas are forced to migrate because of extreme poverty reasons, while 24.6% of the migrants goes to the city in order to improve the living conditions of their households and 25.4% of them (the better-off) departs to accumulate their already achieved prosperity." (Smit 2012: 72)

There is no universal action rationality shaping translocalization strategy in Africa.

This differentiation corresponds to the context-variant orientations for action described in Sect. 4.1 and reinforces our assertion that strategies of migration and translocalization do not follow a universal action rationality.

Greiner (2008, 2011) also highlights the need for a distinguishing view of this aspect:

> Empirical evidence from Namibia suggests that both rich and poor families are involved in migration practices, albeit with different implications. (Greiner 2011: 607).

Better-off households are better-off in the city and in the countryside; poor households are poor in the city and in the countryside.

In his study, he constrasts the accumulative possibilities translocalization holds for wealthier households with the existential necessity which forces resource-poor households into migration. Families that are better off economically are much less dependent on remittances for sustaining their everyday lives; this financial advantage gives the migrated members the chance to invest in their own education, for example, and thus improves their future economic prospects. Moreover, if remittances are made within resource-rich households, this additional capital can be invested at the rural household locations (e.g., in livestock), likely further improving their chances for socioeconomic advancement:

> Rising incomes enable those who earn them to invest in livestock. Remittances provide the essential means to reduce the off-take of animals for consumption. They also reduce the need for emergency sales below market value in case of urgent need of cash, as widely practised among poorer families. As a consequence, livestock ownership by rich families has grown steadily within the last decades, notwithstanding recurrent droughts. (Greiner 2011: 615)

Thus, according to Greiner (2011: 614–615), those families who are wealthy in the countryside are also wealthy in the city, and vice versa. As such, the reverse can be observed in low-resource households: Poor rural households are also poor in the city.

> Poor families [...] are often caught up in survivalist networks, where small incomes are immediately redistributed to satisfy the most basic needs. (Greiner 2011: 614)

The income generated in the city by translocal household members is mainly used for consumer purposes in rural household locations and is only rarely invested toward wealth-building aims such as education.[42] Thus, despite translocalization, poor households find themselves—in sharp contrast to richer households—in a downward socioeconomic spiral (Greiner 2011: 614–616):

> For poorer families, translocality is often a sheer struggle for survival. Sustaining rural and urban family members can be burdensome, and may even be erosive.[43]

Amplification of socioeconomic stratification through translocality

The socioeconomic stratification in the regions of origin and the aggravation of economic disparities through rural–urban migration and remittances that Greiner (2008, 2011) describes are corroborated by Waddington and Sabates-Wheeler (2003: 9–10), Steinbrink (2009, 2010) and Greiner and Sakdapolrak (2012).

Looking at the geographical traits of the translocal ties of rich and poor households, Bleibaum (2009, cited by Jónsson 2010: 11) compares two villages in Senegal's *peanut basin*. He observes that migrations from the more prosperous of the two villages tend to extend over longer distances and involve longer average lengths of stay. In contrast, the migrations from the resource-poorer village usually comprise seasonal movements to nearby cities. Moreover, Reardon (1997) finds that "capital-intensive nonfarm activities tend to be dominated by richer households with greater initial assets (migration remittances, livestock holdings, land)" (Reardon 1997: 741). Studies from Senegal (see Fall and Diagana 1992), Northern Nigeria (Matlon 1979), Burkina Faso and Niger (Reardon et al. 1994) testify to similar patterns.[44]

All in all, it hardly seems possible to say at present whether the observable stratifications and socioeconomic inequalities in rural Sub-Saharan Africa are the triggers or the

---

[42]Lipton (1980) argued similarly as early as 1980; see also Waddington and Sabates-Wheeler (2003) as well as Olson et al. (2004).

[43]"Translocality for these [rich] family networks provides abundant opportunities to redistribute surpluses between rural and urban areas. Indeed, observations from my multi-sited fieldwork suggest that families that are comparatively rich in urban areas are rich in the countryside too. The same can be observed for poor families, with little chance of upward mobility in either rural or urban areas." (Greiner 2011: 615)

[44]A study on the impact of internal and international migration on the position of women left behind in rural Morocco indicates that wives of internal migrants receive fewer remittances and lead economically insecure lives, whereas the wives of international migrants are economically better-off due to higher remittances (de Haas and van Rooij 2010: 54).

outcome of translocalization processes. The contexts, logics of action and migration motives are simply too diverse. There are, however, clear indications that existing disparities within rural communities are exacerbated rather than mitigated by the translocal organization of livelihoods.

5.3.5 Communication Patterns

In Sub-Saharan Africa, too, communication patterns in translocal contexts have been subject to rapid processes of change in recent years. Whereas, (sporadic) letter writing and (ir)regular visits (see Sects. 5.3.2 and 6.1.1.1) long served as the channels for translocal flows of information, over the past 10 years, these have been increasingly supplemented or replaced by new mobile technologies.[45] Due in large part to the poorly developed landline infrastructure,[46], there has been a veritable boom in the use of mobile communications in Africa: The number of mobile phone contracts rose from fewer than 25 million (2001) to approximately 650 million (2012) in just one decade. This figure is higher than in the USA or the European Union, for example. Africa is the fastest growing market for mobile telephony. Today, more than two-thirds of adults in Africa have access to modern information and communication technologies (see Yonazi et al. 2012).

People's increasing connectedness via *telephone and internet has an immense influence on the everyday life and communication practices of both urban and rural populations.*

As people become increasingly connected by telephone and Internet, it has an immense effect on the everyday life and communication practices of both urban and rural populations and thus on the relationship between them. For transmigrants and those "staying home", the improved channels of communication have significantly simplified the previously often challenging task of keeping in touch over

[45]Regarding Hausa migrants in Niger, Youngstedt writes (2013: 148): "Prior to the late 1990s, Hausa separated by migration communicated with each other primarily by proxy; that is, news and cash and gifts were given to individuals traveling back and forth."

[46]"In the year 2000, there were fewer than 20 million fixed-line phones across africa, a number that had accumulated slowly over a century, and a waiting list of a further 3.5 million. With a penetration rate of just over 2%, phones were to be found only in offices and the richest households. But the coming of the mobile phone has transformed communications access." (Yonazi et al. 2012: 13)

long distances. The new technologies have strengthened social networks, while at the same time considerably changing social dynamics within these network structures (Ramisch 2015; Youngstedt 2013: 138, 148; Sward 2016: 25). The rapid developments in the field of mobile telecommunications and the associated technology for mobile money transfer (see Sect. 6.1.1.1) have greatly expanded the possibilities for the translocal organization of livelihoods, enabling members of translocal households who are not present to now play a much greater role in decision-making processes regarding health, education, agriculture, etc. (see Morawczynski 2011; Murphy and Priebe 2011; Ramisch 2015).

> Mobile communication has the potential to transform the resource base of poor people, for example their access to often crucial agriculture, business, education, health and other types of information; the ability to obtain support in situations of psychological stress; the option for increased political participation; and, the benefits from the money and time saved as a result of the better coordination and organization of one's daily activities. In most cases this outweighs the sometimes substantial costs associated with mobile handsets and their usage. (Sterly 2015: 43)

Licoppe (2004) describes how the "*connected presence*" induced by mobile communication both further strengthens "*strong ties*" and creates and expands "*weak ties*", thereby promoting the creation and maintenance of translocal social (security) networks between migrants and people in the region of origin (see also Porter 2012; Shrum et al. 2011; Ramisch 2015; Steel et al. 2017).[47]

Mobile telecommunications technologies are extremely well suited to translocal life.

In view of the fact that an extremely high proportion of the total population in Sub-Saharan Africa is involved in translocal livelihood systems (see Sect. 5.2), we can state beyond a doubt that mobile communications technology is extremely well suited to translocal life. Even before the mobile phone era, communication was the *sine qua non* in translocal contexts. Today, translocal communication is

[47]In addition to the positive side of improved participation opportunities and the maintenance of emotional bonds, this simplified and often very frequent translocal communication sometimes also entails "social stress": Youngstedt (2013), for example, reports that migrants find it more difficult to refuse requests for financial or material support addressed to them directly on the telephone than to do so by letter or email (Youngstedt 2013: 149).

possible almost everywhere and at any time. And this circumstance has an enormous influence on different spheres of people's lives.

The immense growth of the mobile communications market in Africa is itself a sign that the actors embedded in translocal structures regard space- and border-spanning communication as indispensable. This fact becomes even more obvious when one considers how many economically very weak households are prepared to pay an often enormously high proportion of their monthly disposable income for improved communication options: While in Germany, for example, only 0.6% of the average monthly income is needed to operate a mobile phone, in Liberia this figure is 40%, in Malawi 49%, in the Central African Republic 51% and in the Democratic Republic of Congo 52%.[48] Compared to other regions of the world, access to mobile communications is extremely expensive in African countries. However, mobile phones and smartphones are by no means luxury items in Africa; although the use of mobile technology is a costly measure, for many people it is of existential importance to maintain the translocal network relations they rely on. This also explains why in Burkina Faso, for example, it is not unusual for people to have two or three mobile phones operating on different networks, in order to always be able to make calls at the cheapest rate and to compensate for any network disruptions or failures.

Living a translocal life without a mobile phone is hardly imaginable in Africa today.

Mobile telephony has become the most important medium within translocal livelihood systems.[49] In Africa too, a translocal way of life is hardly imaginable without a mobile phone today.

References

ACCRON (2011) Research proposal. African rural-city connections. Work programme: SSH 2010.4.1–2, Connections between rural areas and cities in Sub-Saharan Africa 1–73

ACP (2012) Transnational families and the social and gender impact of mobility in ACP Countries. Background Note: ACPOBS/BN06

---

[48]Calculated based on gross national income per capita (as of 2014) (Brandt 2016).

[49]Youngstedt (2013: 149) finds that 85% of Hausa migrants in Niamey own a mobile phone to keep in touch with family members.

Addai I (2011) Estimating remittances in the informal sector labour market in a developing economy: a micro-level evidence on Kayayoo migrants in Kumasi, Ghana. Soc Sci 6(4):313–317

Adepoju A (2006) Internal and international migration within Africa. In: Kok P, Gelderblom D, Oucho JO, van Zyl J (eds) Migration in South and Southern Africa. HSRC Press, Cape Town, pp 26–45

Agesa RU (2004) One family, two households: rural to urban migration in Kenya. Rev Econ Household 2:161–178

Apeldoorn G (1981) Perspectives on drought and famine in Nigeria. George Allen & Unwin Ltd., London

Ardington C, Case A, Hosegood V (2009) Labor supply responses to large social transfers: longitudinal evidence from South Africa. Am Econ J Appl Econ 1(1):22–48

Awumbila M (2015) Women moving within borders. Gender and internal migration dynamics in Ghana. Ghana J Geogr 7(2):132–145

Awumbila M, Benneh Y, Teye JK, Atiim G (2014) Across artificial borders: an assessment of labour migration in the ECOWAS region. ACP Observatory on Migration, Brussels

Bah M, Cisse S, Diyamett B, Diallo G, Lerise F, Okali D, Okpara E, Olawoye J, Tacoli C (2003) Changing rural-urban linkages in Mali, Nigeria and Tanzania. Environ Urban 15(1):13–23

Baker J, Aina TA (eds) (1995) The migration experience in Africa. Nordiska Afrikainstitutet, Sweden

Beauchemin B (2011) Rural-urban migration in West Africa. Towards a reversal? Migration trends and economic situation in Burkina Faso and Côte d'Ivoire. Popul Space Place 17:47–72

Beauchemin C, Bocquier P (2004) Migration and urbanisation in Francophone West Africa: an overview of the recent empirical evidence. Urban Stud 41(11):2245–2272

Beguy D, Bocquier P, Zulu EM (2010) Circular migration patterns and determinants in Nairobi slum settlements. Demogr Res 23:549–586

Bigsten A (1988) A note on the modelling of circular smallholder migration. Econ Lett 28(1):87–91

Black R, Crush J, Peberdy C, Amassari S, McLean Hilker L, Mouillesseaux S, Pooley C, Rajkotia R (2006) Migration and development in Africa: an overview. Idasa Publishing, Cape Town

Bledsoe C, Ewbank D, Isiugo-Abanihe U (1988) The effect of child fostering on feeding practices and access to health services in rural Sierra Leone. Soc Sci Med 27(6):627–636

Bleibaum F (2009) Senegal case study report. EACH-FOR environmental change and forced migration scenarios. Available via EACH-For. http://rosamartinez.org/wp–content/uploads/2015/11/Migraciones–y–Cambio–Climatico_EACHFOR.pdf. Accessed 05 Mar 2019

Bocquier P (2003) Analyzing urbanization in Africa. In: Hugo G, Champion A (eds) New forms of urbanisation. Ashgate, Aldershot, pp 133–150

Bocquier P, Traoré S (eds) (2000) Urbanisation et dynamique migratoire en Afrique de l'Ouest—la croissance urbaine en panne. L'Harmattan, Paris

Brandt M (2016) Mobilfunk: in Deutschland billig, in Afrika Luxus. Available via Statista. https://de.statista.com/infografik/4362/anteil-mobilfunkkosten-am-einkommen/. Accessed 22 Mar 2019

Brockerhoff M, Eu H (1993) Demographic and socioeconomic determinants of female rural to urban migration in Sub-Saharan Africa. Int Migrat Rev 27(3):557–577

BUCREP (Bureau Central des Recensements et des Etudes de Population) (2011) Rapport National sur l'etat de la population. Available via bucrep. http://www.bucrep.cm/index.php/fr/ressources-et-documentations/telechargement/category/44-rapport-national-sur-l-etat-de-la-population. Accessed 18 Mar 2019

Castaldo A, Deshingkar P, McKay A (2012) Internal migration, remittances and poverty: evidence from Ghana and India. University of Sussex. Available via Migrating out of Poverty. http://migratingoutofpoverty.dfid.gov.uk/files/file.php?name=wp7-internal-migration-remittances-and-poverty.pdf&site=354. Accessed 18 Mar 2019

Chant S (1998) Households, gender and rural-urban migration. Reflections on linkages and considerations for policy. Environ Urban 10(1):5–21

Clark SJ, Collinson MA, Kahn K, Drullinger K, Tollman SM (2007) Returning home to die: circular labour migration and mortality in South Africa. Scand J Public Health 69:35–44

Clouse K, Fox MP, Mongwenyana C, Motlhatledi M, Buthelezi S, Bokaba D, Norris SA, Bassett J, Lurie MN, Aronoff DM, Sten H (2018) 'I will leave the baby with my mother': long-distance travel and follow-up care among HIV-positive pregnant and postpartum women in South Africa. J Int AIDS Soc 21:74–82

Collinson M, Tollman S, Kahn K (2007) Migration settlement change and health in post-apartheid South Africa: triangulating health and demographic surveillance with national census data. Scand J Public Health 69:77–84

Collinson M, Tollman S, Kahn K, Clark S (2003) Highly prevalent circular migration: households, mobility and economic status in rural South Africa. University of Witwatersrand

Cordell DD (1996) Hoe and wage: a social history of a circular migration system in West Africa (African Modernization and Development Series). Westview Press, Boulder

Cottyn I, Schapendonk J (2013) Mobility in Sub-Saharan Africa: patterns, processes and policies. RurbanAfrica: State of the art report 2, Department of Geosciences and Natural Resource Management, University of Copenhagen

Cremin I, Morales F, Jewell B, O'Reilly KR, Hallett TB (2015) Seasonal PrEP for partners of migrant miners in Southern Mozambique: a highly focused PrEP intervention. J Int AIDS Soc 18:1–7

Crush J, Williams V, Peperdy S (2005) Migration in Southern Africa. Available via Global Commission on international Migration, University of Cape Town/ Queens University. https://womin.org.za/images/impact-of-extractive-industries/migrancy-and-extractivism/Crush%20%20Williams%20and%20Peberdy%20-%20Migration%20in%20Southern%20Africa.pdf. Accessed 18 Mar 2019

De Haas H (2010) Migration and development: a theoretical perspective. Int Migrat Rev 1(44):227–264

De Jong F (1999) The production of translocality. In: De John F (ed) Modernity on a shoestring: dimensions of globalization, consumption and development in Africa and beyond. Based on an EIDOS conference held at The Hague, 13–16 March 1997. EIDOS, Leiden, pp 315–340

Deshingkar P (2012) Environmental risk resilience and migration: implications for natural resource management and agriculture. Environ Res Lett 7:1–7

Deshingkar P, Wood RG, Béné C (2015) Adaptive social protection and migration. Final report. Available via Institute of Development Studies, Brighton. https://assets.publishing.service.gov.uk/media/57a0897ced915d3cfd000280/61459-Adaptive_Social_Protection_and_Migration_Report.pdf. Accessed 18 Mar 2019

Devereux S (2006) Vulnerable livelihoods in Somali region, Ethiopia. Brighton, Institute of Development Studies

Doevenspeck M (2005) Migration im ländlichen Benin: Sozialgeographische Untersuchungen an einer afrikanischen Frontier. Verlag für Entwicklungspolitik, Saarbrücken

Doevenspeck M (2011) The thin line between choice and flight: environment and migration in rural Benin. Int Migr 49:50–68

Dorlöchter-Sulser S (2014) Wandel von Livelihood-Systemen im Spannungsfeld von Struktur und Handeln, eine historisch angelegte Livelihood-Analyse von 1960 bis 2010 am Beispiel der Region Dosso. Niger, Wvb, Berlin

Douglass M (1998) A regional network strategy for reciprocal rural-urban linkages: an agenda for policy research with reference to Indonesia. Third World Plan Rev 20(1):24–47

Elkan W (1967) Circular migration and the growth of towns in East Africa. Int Labour Rev 96:581–589

Etzold B (2017) Mobility, space and livelihood trajectories: new perspectives on migration, translocality and place-making for livelihood studies. In: De Haan L (ed) Livelihoods and development: new perspectives. Brill, Leiden, pp 44–68

Fall A, Diagana B (1992) Les activités non agricoles et la capacité de financement de la consummation alimentaire par les ménages ruraux au Sénégal. Paper presented at PRISAS/Institut du Sahel Workshop on non-agricultural informal activities and food security in the Sahel, PRISAS/Institut du Sahel, Bamako Mali, 12–14 May 1992

Fischer C, Vollmer R (eds) (2009) Migration and displacement in Sub-Saharan Africa: the security-migration nexus. Bonn International Center for Conversion, Bonn

Forsberg J (1999) Provisioning by rural-urban interaction and mobility adjusted livelihoods and social relations in Mvumi, Tanzania. In: Paper presented at the conference 'between town & country livelihoods settlement and identity formation in Sub-Saharan Africa', Rhodes University, East London, 27–30 June 1999

Gandhi D (2018) Figures of the week: internal migration in Africa. Available via Brookings. https://www.brookings.edu/blog/africa-in-focus/2018/06/07/figures-of-the-week-internal-migration-in-africa/. Accessed 16 Apr 2016

Gibson MA, Gurmu E (2012) Rural to urban migration: is an unforeseen impact of development intervention in Ethiopia. Available via PLoS ONE 7(11):e48708. https://doi.org/10.1371/journal.pone.0048708. Accessed 18 Mar 2019

Giesbert L (2007) Seeking opportunities migration as an income diversification strategy of households in Kakamega District in Kenya. German Institute of Global and Area Studies, Hamburg

Gilbert AG (1992) Third world cities housing infrastructure and servicing. Urban Stud 29:435–460

Gould WTS (1989) Technical education and migration in Tiriki, Western Kenya 1902–1987. Afr Aff 88(351):253–271

Grawert E (1998) Der 'livelihood approach': Eine Analysemethode für komplexe sozioökonomische Absicherungsformen dargestellt am Beispiel Westsudan. Peripherie 69:67–87

Gregoire E (1992) The Alhazi of Maradi traditional hausa merchants in a changing Sahelian City. Lynne Rienner, Boulder

Greiner C (2008) Zwischen Ziegenkraal und Township: Migrations prozesse in Nordwest-Namibia. Dietrich Reimer, Berlin

Greiner C (2010) Patterns of translocality: migration, livelihoods and identities in Northwest Namibia. Sociologus 2(60):131–161

Greiner C (2011) Migration, translocal networks and socio-economic stratification in Namibia. Africa 81(4):606–627

Greiner C, Sakdapolrak P (2012) Rural-urban migration, agrarian change, and the environment in Kenya: a critical review of the literature. Popul Environ 34(4):524–553

Gugler J (1982) Overurbanization reconsidered. Econ Dev Cult Change 31(1):173–189

Gugler J, Ludwar-Ene G (1995) Gender and migration in Africa South of the Sahara. In: Baker J, Aina TA (eds) The migration experience in Africa. Nordiska Afrikainstitutet, Sweden, pp 257–269

Güneralp B, Lwasa S, Masundire H, Parnell S, Karen CS (2018) Urbanization in Africa: challenges and opportunities for conservation. Environ Res Lett 13. Available via IOP Publishing Ltd. https://iopscience.iop.org/article/10.1088/1748-9326/aa94fe/pdf. Accessed 16 Apr 2019

Hahn HP (2004) Zirkuläre Arbeitsmigration in Westafrika und die 'Kultur der Migration'. Afr Spectr 39:381–404

Hashim I (2007) Independent child migration and education in Ghana. Dev Change 38(5):911–931

Haug R (2002) Forced migration, processes of return and livelihood construction among pastoralists in northern Sudan. Disasters 26 (1):70–84

Hazell PBR, Hojjati B (1995) Farm-non-farm growth linkages in Zambia. J Afr Econ 4(3):406–435

Hoddinott J (1994) A model of migration and remittances applied to Western Kenya. Oxford Econ Papers 46(3):459–476

IIED International Institute for Environment and Development (2001a) Potentialités et conflits dans les zones péri-urbaines: le cas de Bamako au Mali. Working Paper 5, Series on Rural-Urban Interactions and Livelihood Strategies. Available via IIED. https://pubs.iied.org/pdfs/9119IIED.pdf. Accessed 16 Apr 2019

IIED International Institute for Environment and Development (2001b): Potentialités et conflits dans les zones péri-urbaines: le cas de Mopti au Mali. Working paper 6, Series on rural-urban interactions and livelihood strategies. Available via IIED

IOM (2013) Migration and development within the south: new evidence from African Caribbean and Pacific countries. In: International organization for migration-migration research series 46. Available via IOM. https://publications.iom.int/books/migration-and-development-within-south-new-evidence-african-caribbean-and-pacific-countries. Accessed 18 Mar 2019

Isiugo-Abanihe UC (1985) Child fosterage in West Africa. Popul Dev Rev 11(1):53–73

Johnson GE, Whitelaw WE (1974) Urban-rural income transfers in Kenya: an estimated-remittances function. Econ Dev Cult Change 22 (3):473–479

Jónsson G (2010) The environmental factor in migration dynamics—a review of African case studies. In: International migration institute working paper 21, University of Oxford

Klocke-Daffa S (2015) Hauskinder—Großmachkinder—Logierkinder: Die Aufnahme von Kindern als Ressource im Kontext sozialer Sicherung. In: Fink M, Gronemeyer R (eds) Afrikanische Kindheiten. Soziale Elternschaft und Waisenhilfe in der Subsahara, Transcript, Bielefeld, pp 59–78

Kok P, Collinson M (2006) Migration and urbanization in South Africa. Statistics South Africa Report 03-04-02, Pretoria

Krüger F (1998) Taking advantage of rural assets as a coping strategy for the urban poor: the case of rural-urban interrelations in Botswana. Environ Urban 10(1):119–134

Kwankye SO, Anarfi JK, Tagoe CA, Castaldo A (2009) Independent north-south child migration in Ghana: The decision-making process. Working Paper T-29, Development and Research Centre on Migration, Globalisation and Poverty, Sussex

Lachaud J, LeGrand TK, Kobiané JF (2016) Child fostering and children's human capital in Ouagadougou. Popul Rev 55(1):27–48

Lerise F, Kibadu A, Mbutolwe E, Mushi N (2001) The case of Lindi and its region Southern Tanzania IIED. Briefing paper series on rural-urban interactions and livelihood strategies, Briefing Paper 2

Licoppe C (2004) 'Connected' presence: the emergence of a new repertoire for managing social relationships in a changing communication technoscape. Environ Plan D Soc Space 22(1):135–156

Lipton M (1980) Migration from rural areas of poor countries: the impact on rural productivity and income distribution. World Dev 8 (1):1–24

Lohnert B (2002) Vom Hüttendorf zur Eigenheimsiedlung: Selbsthilfe im städtischen Wohnungsbau: Ist Kapstadt das Modell für das neue Südafrika?. Universitätsverlag Rasch, Osnabrück

Lohnert B, Steinbrink M (2005) Rural and urban livelihoods: a translocal perspective. S Afr Geogr J 87:95–103

Martin J (2015) "Ein Kind gehört dir nicht allein." Soziale Elternschaft in Nordbenin. In: Fink M, Gronemeyer R (eds) Afrikanische Kindheiten. Soziale Elternschaft und Waisenhilfe in der Subsahara. Transcript, Bielefeld, pp 43–57

Matlon P (1979) Income distribution among farmers in Northern Nigeria: empirical results and policy implications. In: African rural economy paper no 18, Michigan State University

Mberu B (2005) Who moves and who stays? Rural out-migration in Nigeria. J Popul Res 22(2):141–161

Mberu B, Langat GC, Kimani J, Oti SO (2013) Family ties and urban-rural linkages among older migrants in Nairobi informal settlements. Popul Space Place 19:275–293

Morawczynski O (2011) Examining the adoption usage and outcomes of mobile money services: the case of M-PESA in Kenya. PhD thesis. Science & Tech Studies, Edinburgh University

Morrissey JW (2013) Understanding the relationship between environmental change and migration: the development of an effects framework based on the case of northern Ethiopia. Glob Environ Change 23(6):1501–1510

Murphy L, Priebe AE (2011) 'My co-wife can borrow my mobile phone!' gendered geographies of cell phone usage and significance for rural Kenyans. Gender Technol Dev 15(1):1–23

Muzvidziwa VN (1997) Rural-urban linkages Masvingo's double rooted female heads of households. Zambezia XXIV(6):97–123

Newland K, Rannveig Agunias D, Terrazas A (2009) Circular migration and human development. United Nations Development Programme, Research Paper 42

Obeng-Odoom F (2010) Abnormal urbanization in Africa: a dissenting view. Afr Geogr Rev 29(2):13–40

Ogude HR, de Berry JP, Sarkar A, Baum C (2018) Mixed migration, forced displacement and job outcomes in South Africa. Available via The World Bank Group. http://documents.worldbank.org/curated/en/247261530129173904/pdf/127706-v2-WP-P163910-main-MixedMigration-ForcedDisplacement-and-Job-Outcomes-in-South-Africa-Final-June-2018-PUBLIC-Volume-2-1.pdf. Accessed 18 Mar 2019

Okali D, Okpara E, Olawoye J (2001) The case of Aba and its region, Southeastern Nigeria. In: Working paper series on rural-urban interactions and livelihood strategies. Working Paper 4, International Institute for Environment and Development

Olson JM, Atieno F, Muchugu E (2004) Multi-scale analysis of land use and management change on the eastern slopes of Mt. Kenya. In: LUCID project working paper 20, Michigan State University

Oltmer J (2015) Zusammenhänge zwischen Migration und Entwicklung. Available via Deutsche Welthungerhilfe e. V. https://www.welthungerhilfe.de/aktuelles/publikation/detail/zusammenhaenge-zwischen-migration-und-entwicklung/. Accessed 18 Mar 2019

Ouédraogo D (2009) Migration and population in West Africa: Political issues and perspectives. In: Trémolières M (ed) Regional challenges of West African migration—African and European perspectives. Available via OECD. http://www.oecd.org/publications/regional-challenges-of-west-african-migration-9789264056015-en.htm. Accessed 18 Mar 2019

Ouédraogo VS (2012) Von Côte d'Ivoire nach Burkina Faso: Re-Migration und Integration—Die Perspektive burkinischer Arbeitsmigrantinnen und -migranten. Logos, Berlin

Paterson DB (1979) Household resource allocation among the Luhya of east Bunyore: a case study approach. In: Working paper no. 351, Nairobi: Institute for Development Studies, University of Nairobi

Pilon M (2003) Confiage Et Scolarisation En Afrique De L'Ouest: Un Etat Des Connaissances. Available via ISSP. http://www.issp.bf/index.php/fr/droits-des-enfants-au-bf/2-documents-recenses/1-education/4-rapport-de-recherche-consultation/116-confiage-et-scolarisation-en-afrique-de-l-ouest/file. Accessed 22 Mar 2019

Piot C (1999) Remotely global village modernity in West Africa. University of Chicago Press, Chicago

Porter G (2012) Mobile phones, livelihoods and the poor in Subsaharan Africa. Geogr Compass 6(5):241–259
Posel D (2003) Have migration patterns in Post-Apartheid South Africa changed? In: Paper prepared for Conference on African migration in comparative perspective, Johannesburg, 4–7 June 2003
Potts D (2009) The slowing of Sub-Saharan Africa's urbanization: evidence and implications for urban livelihoods. Environ Urban 21 (1):253–259
Potts D (2010) Circular migration in Zimbabwe & contemporary Sub-Saharan Africa. James Currey, Woodbridge
Potts D (2015) Debates about African urbanization migration and economic growth what can we learn from Zimbabwe and Zambia? Geogr J 182(3):1–14
Potts D, Mutambirwa C (1990) Rural-urban linkages in contemporary Harare: Why migrants need their land. J South Afr Stud 16(4):677–698
Pursey C (2014) Rural-urban mobilities and belonging in the Bamilike and Peul Communities of Yaoundé, Cameroon. Unpublished MSc research report. Utrecht University, Utrecht
Rain D (1999) Eaters of the dry season: circular labor migration in the West African Sahel. Westview Press, Boulder
Ramisch J (2014) 'We will not Farm Like our Fathers did' multilocational livelihoods cellphones and the continuing challenge of rural development in Western Kenya. In: Sick D (ed) Rural livelihoods regional economies and processes of change. Routledge, London, pp 10–25
Ramisch J (2015) 'Never at Ease': cellphones multilocational livelihoods and the metabolic rift in Western Kenya. Agric Hum Values 33(4):979–995
Ratha D, Mohapatra S, Özden C, Plaza S, Shaw W, Shimeles A (2011) Leveraging migration for Africa: remittances skills and investments. Available via World Bank Group. https://siteresources.worldbank.org/EXTDECPROSPECTS/Resources/476882-1157133580628/AfricaStudyEntireBook.pdf. Accessed 18 Mar 2019
Rauch T, Beckmann G, Neubert S, Rettberg S (2016) Ländlicher Strukturwandel in Subsahara Afrika. SLE discussion paper 01/2016. Available via SLE. https://www.sle-berlin.de/files/SLE%20Discussion%20Paper%2001-Strukturwandel%20mit%20Cover.pdf. Accessed 18 Mar 2019
Ravenstein EG (1885) The laws of migration. J Stat Soc 48:167–227
Reardon T (1997) Using evidence of household income diversification to inform study of the non-farm labour market in Africa. World Dev 25(5):735–748
Reardon T, Fall AA, Kelly V, Delgado C, Matlon P, Hopkins J, Badiane O (1994) Is income diversification agriculture-led in the West African semi-arid tropics? The nature causes effects distribution and production linkages of off-farm activities. In: Atsain A, Wangwe S, Drabek AG (eds) Economic policy experience in Africa: what have we learned?. African Economic Research Consortium, Nairobi, pp 207–230
Reed HE, Andrzejewski CS, White MJ (2010) Men's and women's migration in Coastal Ghana: an event history analysis. Demogr Res 22(25):771–812
Ross MH, Weisner TS (1977) The rural-urban migrant network in Kenya: some general implications. Am Ethnol 4(2):359–375

Saghir J, Santoro J (2018) Urbanization in Sub-Saharan Africa. Meeting challenges by bridging stakeholders. Available via Center for Strategic & International Studies. https://csis-prod.s3.amazonaws.com/s3fs-public/publication/180411_Saghir_UrbanizationAfrica_Web.pdf. Accessed 16 Apr 2019

Schmidt-Kallert E (2009) A new paradigm of urban transition: tracing the livelihood strategies of multi-locational households. Erde 140:319–336

Schutten M (2012) Rwanda livelihood diversity: causes of rural-urban migration; Why Rwanda poverty classification does not explain migration flows. Master Dissertation, University of Utrecht and National University of Rwanda

Shrum WPN, Mbatia A, Palackal D-BS, Duque Dzorgbo RB, Ynalvez MA (2011) Mobile phones and core network growth in Kenya: strengthening weak ties. Soc Sci Res 40(2):614–625

Smit A (2012) Impacts of rural-urban migration on rural migrant households in the surroundings of Kigali. Master Thesis, University of Utrecht

Sporton D, Thomas DSG, Morrison J (1999) Outcomes of social and environmental change in the Kalahari of Botswana: the role of migration. J South Afr Stud 15(3):441–459

Steel G, Cottyn I, van Lindert P (2017) New connections—new dependencies: spatial and digital flows in Sub-Saharan African livelihoods. In: De Haan L (ed) Livelihoods and development: new perspectives. Brill, Leiden, pp 148–167

Steinbrink M (2009) Leben zwischen Land und Stadt, Migration. Translokalität und Verwundbarkeit in Südafrika, Springer, Wiesbaden

Steinbrink M (2012) Migration, Netzwerk und 'Entwicklung'. Translokale Livelihoods in Südafrika. IMIS-Beiträge 42:165–204

Sterly H (2015) 'Without a Mobile Phone I Suppose I Had to Go There'—mobile communication and translocal social constellations in Bangladesh. ASIEN 134(1):31–46

Sward J (2016) Moving to 'Greener Pastures'? The complex relationship between internal migration, land tenure and poverty in mid-Ghana. In: RPC working paper no. 33. Migrating out of Poverty Consortium, University of Sussex, Brighton

Tacoli C (1998) Rural-urban interactions: a guide to the literature. Environ Urban 10(1):147–166

Tacoli C (2002) The links between urban and rural development. Environ Urban 15(3):3–12

Tacoli C, Mabala R (2010) Exploring mobility and migration in the context of rural-urban linkages: why gender and generation matter. Environ Urban 22(2):389–395

Tamanja EMJ (2014) Child migration and educational progression in the Savannah regions of Ghana: implications for planning and spatial development. Dissertation, University of Technology, Dortmund

Taylor-Powell E (1992) Social soundness analysis: disaster preparation: mobile technology and politics in rural West Bengal, India. Ethos 7 (3):398–420

Todaro M (1969) A model of labour migration and urban unemployment in less developed countries. Am Econ Rev 59:138–148

Tostensen A (1986) Between shamba and factory: preliminary results from a study of oscillatory labour migration in Kenya. University of Nairobi, Institute for Development Studies, Working Paper 423

Toure M 1998 Country Report Côte d'Ivoire. In: Paper presented at the regional meeting Trends and Prospects for the 21st Century, Gaborone, 2–5 June 1998

Trager L (1998) Home-town linkages and local development in South West Nigeria: whose agenda? What impact? J Int Afr Inst 68 (3):360–382

Tunali I (2000) Rationality of migration. Int Econ Rev 41(4):893–920

UN (2018) Economic development in Africa report. Migration for structural transformation. New York, Geneva

UN Population Division (2015) World urbanization prospects. The 2014 revision. Available via DESA. https://esa.un.org/unpd/wup/publications/files/wup2014-report.pdf. Accessed 22 Mar 2019

UN Population Fund (2010) State of world population. UNFPA. https://www.unfpa.org/publications/state-world-population-2010. Accessed 18 Mar 2019

UNDP (2014) Human development report. Sustaining human progress: reducing vulnerabilities and building resilience. Available via UNDP. http://hdr.undp.org/sites/default/files/hdr14-report-en-1.pdf. Accessed 22 Mar 2019

van Westen ACM (1995) Unsettled: low-income housing and mobility in Bamako, Mali. Royal Dutch Geographical Society/Faculteit Ruimtelijke Wetenschappen Universiteit Utrecht, Utrecht

Vorlaufer K (1992) Wanderungen zwischen ländlichen Peripherie- und großstädtischen Zentralräumen. Zeitschrift für Wirtschaftsgeographie 28(4):229–261

Waddington H, Sabates-Wheeler R (2003) How does poverty affect migration choice? A review of literature. In: IDS working paper T3. Brighton

Werthmann K (2007) Dans un monde masculin le travail de femmes dans un camp de chercheurs d'or au Burkina Faso. In: Boesen E, Marfaing L (eds) Les nouveaux urbains dans l'espace Sahara-Sahel. Karthala, Paris, pp 295–322

World Bank (2016) Population, total. Available via The World Bank. https://data.worldbank.org/indicator/SP.POP.TOTL. Accessed 18 Mar 2019

Yonazi E, Kelly T, Halewood N, Blackman C (eds) (2012) The transformational use of information and communication technologies in Africa. eTransform report Africa. Available via The World Bank. http://web.worldbank.org/WBSITE/EXTERNAL/TOPICS/EXTINFORMATIONANDCOMMUNICATIONANDTECHNOLOGIES/0,contentMDK:23262578~pagePK:210058~piPK:210062~theSitePK:282823,00.html. Accessed 18 Mar 2019

Youngstedt SM (2013) Voluntary involuntary homebodies adaptations and lived experiences of Hausa 'left behind' in Niamey, Niger. In: Kane A, Leedy TH (eds) African migrations patterns and perspectives. Indiana University Press, Bloomington, Indianapolis, pp 133–157

Zelinsky W (1971) The hypothesis of the mobility transition. Geogr Rev 64(3):219–249

Ziegelmayer U, Spaan E (2019) Migrant trajectories within the context of demographic, socio-economic and environmental change: evidence from coastal Ghana. In: Hillmann F, Van Naerssen T, Spaan E (eds) Trajectories and imaginaries in migration: the migrant actor in transnational space. Routledge, London, New York, pp 34–55

Chapter 6
Effects of Translocal Livelihoods on Rural Change

Abstract In this chapter, we discuss different dimensions of the effect of translocal livelihoods on rural structural change. Based on a literature review, the chapter is roughly structured according to the three dimensions of sustainability. Section 6.1 ("*Economic Dimension*") focuses first on the effects of remittances and their use, second on the division of labor as well as the distribution of workforce and workload and third on innovation and the knowledge flow in translocal contexts. Section 6.2 ("*Ecological Dimension*") centers around the reciprocal relationship between translocality and changing the natural environment. Here, we argue for conceptually combining the *socio-ecological systems* approach with the concept of translocality ("*translocal socio-ecological systems*"), which would enable researchers not only to consider the effects of environmental change on migration (and translocalization) but also to analyze the feedback effects of translocality on the environment. Section 6.3 ("*Social Dimension*") primarily spotlights *education*, *gender* and *health*, and on the questions, (1) *to what degree* these central social aspects are connected to migration and the translocality of livelihoods, and (2) whether we can discern a shift in values in these areas due to translocality. In sum, this chapter highlights how the economic, ecological and social dynamics of (rural) change are deeply interwoven in the context of translocal livelihoods.

Keywords Rural change · Economic development · Remittances · Labor · Knowledge transfer · Innovation · Natural environment · Education · Gender · Health

In this chapter, we will discuss the different effects of translocal livelihoods on rural structural change. The chapter is roughly structured according to the three dimensions of

M. Steinbrink and H. Niedenführ, *Africa on the Move*, Springer Geography,
https://doi.org/10.1007/978-3-030-22841-5_6

sustainability: "the economic" (Sect. 6.1), "the ecological" (Sect. 6.2) and "the social" (Sect. 6.3).

6.1 Economic Dimensions of Translocal Livelihoods

When assessing the economic dimensions of how translocal livelihoods impact rural structural change in sub-Saharan Africa, we must first recognize that the translocalization process has primarily economic causes. Strong population growth coupled with the largely unfavorable market conditions for agricultural products and simultaneous lack of nonagricultural employment opportunities in rural areas has led to a rural existential crisis in large parts of Africa, making labor migration economically imperative. Because of the precarious labor market conditions in (urban) areas of arrival described in Sect. 5.1, a large proportion of households—often motivated by their dire economic circumstances—organize their livelihoods across several locations, economically diversifying them by combining agricultural subsistence and market production with (often informal) wage labor.

6.1.1 Remittances

In translocal livelihood systems, migrants' monetary remittances to their rural family members are fundamental:

The hope of remittances is the main trigger for translocalization.

The hope of being able to financially support rural household members is widely regarded as the most important trigger for labor migration and translocalization.[1] These transfer activities within translocal households are an essential feature of translocally diversified livelihood systems. Accordingly, this aspect has been well studied.

In order to assess the significance of remittances for processes of rural structural change, the following chapter will first assess, based on existing case studies, the quantitative

[1] A policy letter from the Migrating out of Poverty Research Programme Consortium (2016a: 3) on rural–urban migration in Ghana reports that "even when the rural–urban migrant himself or herself felt migration had not been very helpful in moving him/her out of poverty, sending remittances home is still a top priority."

extent of these remittances and their share in the available income of the rural population.

For some time now, researchers have repeatedly highlighted the outstanding importance of remittances for the economic development of many countries in the Global South (see, among others, Ambrosius et al. 2008: 1). Emphasis is placed on the fact that the international remittances of migrants in sum far exceed the value of the total transfers made worldwide within the framework of Official Development Assistance (ODA).[2] For example, in 2010, the Organization for Economic Cooperation and Development (OECD) announced that the annual ODA of OECD member states had reached a historic high of US$129 billion. According to World Bank estimates, worldwide remittances of migrants in the same year amounted to more than US$440 billion.[3] For Africa, too, international remittances play a major macroeconomic role, albeit with strong regional differences. Table 6.1 shows the magnitude of international remittance flows for various African countries in US dollars and as a share of the countries' respective gross domestic product (GDP). The list shows the enormous significance of international remittances for some African economies. For instance, remittances to Senegal in 2012 totaled US$1,366.85 million and accounted for 11.43% of GDP. In Mali, remittances totaled US$444.45 million, 5.02% of GDP; in Côte d'Ivoire, which has more immigration than emigration, the US$325.09 million in remittances still made up 1.63% of GDP; Togo's US$320.71 million in remittances amounted to 10.61% of GDP; and in Liberia, the payments totaled US$373.39 million, representing a striking 23.41% of GDP (Watkins and Quattri 2014: 11). A strong uptrend can be observed in remittances as well: Whereas US $14 million were transferred to Niger in 2000, for example, that figure was over US$66 million in 2005 and US$78 million in 2007/2008 (Youngstedt 2013: 150).[4]

[2]In 2003, the Global Development Finance Report (see World Bank 2003) reported that for the first time the total volume of migrant remittances exceeded the total volume of transfers made worldwide as Official Development Assistance (ODA); in some countries they were even higher than foreign direct investment (FDI) (see World Bank 2018).

[3]See also: OECD (2010), MPI (2012), Ratha and Riedberg (2005), IOM (2010), OECD (2009). For current figures, see World Bank (2018).

[4]In 2018, the following African countries had the highest shares of remittances relative to GDP: Liberia (27%), Comores (21%), and the Gambia (21%) (see World Bank 2018: 5).

Table 6.1 International remittance flows to sub-Saharan Africa

Millions USD		Share of GDP in % (available data 2010–2012)	
Nigeria	20.568	Liberia	23.41
Senegal	1.366	Lesotho	22.64
Kenya	1.227	Gambia, The	15.37
Sudan	1.126	Senegal	11.43
Uganda	977	Togo	10.61
Lesotho	602	Cabo Verde	9.13
Ethiopia	524	Nigeria	7.86
Mali	444	Guinea-Bissau	5.49
Liberia	372	Mali	5.02
Côte d'Ivoire	325	Uganda	3.69
Togo	321	Kenya	2.98
Mauritius	247	Rwanda	2.57
Benin	179	São Tomé and Principe	2.41
Cape Verde	176	Benin	2.36
Rwanda	156	Niger	2.35
Ghana	152	Burundi	1.88
Burkina Faso	130	Côte d'Ivoire	1.63
Niger	122	Sierra Leone	1.61
Cameroon	109	Mozambique	1.55
Mozambique	99	Ethiopia	1.50
Gambia, The	89	Burkina Faso	1.31
Sierra Leone	79	Guinea	1.18
Tanzania	75	Swaziland	0.84
Guinea	75	Cameroon	0.83
Botswana	55	Sudan	0.68
Swaziland	47	Malawi	0.66
Zambia	46	Zambia	0.35
Guinea-Bissau	42	Ghana	0.34
Burundi	42	Tanzania	0.24
Seychelles	26	Botswana	0.13
Namibia	17	Namibia	0.12
Malawi	16	Seychelles	0.11
São Tomé and Principe	7	Mauritius	0.01
Angola	0.19	Angola	0.00

Source Watkins and Quattri (2014: 11); based on: World Bank (2012); World Bank (2014)

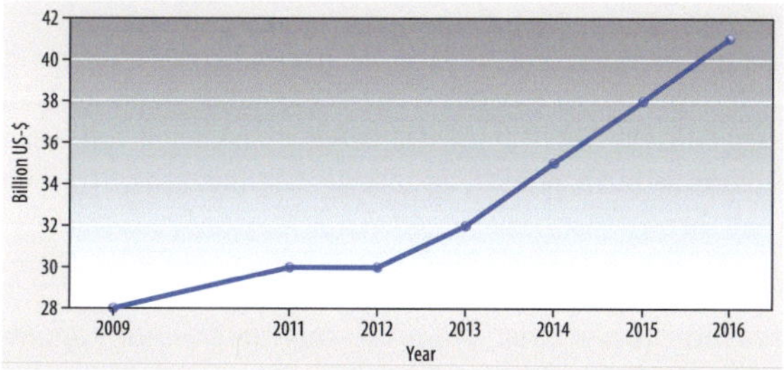

Fig. 6.1 Rising trend: international remittance flows to sub-Saharan Africa (in Bil. US$). *Source* Modified from Watkins and Quattri (2014: 10), from World Bank (2013)

The World Bank sees this sharp increase as part of a more general trend in sub-Saharan Africa and assumes that the volume of international remittances will continue to increase significantly in the coming years (see Fig. 6.1).[5]

The official statistics on international remittances presented here are probably much lower than the actual figures, as a large part of the money is transferred through informal channels. Studies from Burkina Faso and Senegal, for example, have shown that over 60% of recipient households utilize informal channels for cross-border remittances (see Scharwatt and Williamson 2015: 6).

Even more difficult than estimating international money transfers is assessing the remittances from internal migrants. There is no specific and aggregated data on the extent of these flows of remittances in sub-Saharan Africa (see Ayana Aga et al. 2014; Sander and Munzele Maimbo 2005). Information on such transfers could only feasibly be collected in household surveys (Sander and Munzele Maimbo 2005: 6). Thus, while international remittances are presently receiving a great deal of attention in macroeconomic assessments, and efforts are being made to quantify these capital flows, significantly less consideration is being given to the remittances from internal (rural–urban or rural–rural) migrants within any one country:

[5]However, the growth rate of international remittances to developing countries declined from 3.2% in 2014 to just 0.4% in 2015. Probable reasons are economic weakness in the destination countries and a low oil price (KNOMAD 2016: iii).

> Estimates of the total volume of internal remittances are very rare for individual countries and nonexistent on a global level. [...] Furthermore, transfers sent by internal migrants are more often sent through informal channels, making it difficult to capture them in official estimates of remittances. (Castaldo et al. 2012: 16)

Remittances from internal migrants reach far poorer rural households than international remittances.

This deficit of data stands in striking disproportion to the many indications that internal migrants' remittances are existentially important to far poorer rural households than international remittances are. Although the individual sums of money transferred as *domestic remittances* are indeed much smaller on average than international remittances, they come more regularly; and as various case studies suggest, they are greater in their totality than international remittances (Sander and Munzele Maimbo 2005).[6] When comparing domestic and international remittances in Ghana, Cottyn and Schapendonk (2013: 28) detect the following pattern in the frequency, sum, and origin of remittances: While the frequency and the volume of remittances sent within Ghana exceed those from abroad, the dollar amounts transferred per domestic remittance are considerably smaller than the international ones. In concrete terms, 81.8% of receiving households in Ghana are sent domestic remittances and 18.2% receive international transfers; the difference in volume is US $324 million domestic to US$283 million international remittances. This explains the relative balance in the total volume of remittances (53.3% internal, 46.7% international). For Rwanda, South Africa and Uganda too, McKay and Deshingkar (2014) confirm that *domestic remittances* significantly exceed international remittances in volume.

In their comparative study, McKay and Deshingkar (2014: 21) come to the general conclusion that poor households tend to receive remittances from internal migrants, whereas wealthier households benefit more often from international remittances:

> Domestic remittances are more likely to be received by poorer households, while international remittances tend to be received by richer households. If a poor household does

[6]"As for domestic remittance transfers, nowhere in Africa are these captured as a separate category of domestic transactions; they can be traced only through household surveys." (Sander and Munzele Maimbo 2005: 5)

> receive an international remittance, this can have a substantial poverty reduction impact for that household, but few poor households benefit from such remittances. Hence, the bulk of the poverty reduction impact of remittances in all the countries discussed here comes from domestic transfers. (McKay and Deshingkar 2014: 21)

International migration is predominantly a strategy of wealthier households.

This finding alone suggests that wealthier households are more likely to pursue the strategy of international migration; since they do not find themselves in a situation of heightened baseline vulnerability, they have the freedom to set objectives toward benefit maximization (see Sects. 4.2 and 6.1.1.3).

The quantitative data on the remittances of internal migrants, compiled from various case studies and presented in Table 6.2, indicates how many people in rural areas in sub-Saharan Africa receive remittances, offering insight into the extent and frequency of translocal capital flows.

The following two diagrams (Figs. 6.2 and 6.3) illustrate the proportion of different income types in the total monetary income of the resident population in two rural communities in southern Africa (South Africa and Namibia).

While the statistics presented in this chapter show that the significance of domestic remittances varies (at times starkly) between and within African countries, we can glean from present research that internal migrants' transfers are generally of great importance. Given the large proportion of the sub-Saharan population involved in translocal contexts (see Sect. 5.2), the extent of remittance practices should not come as a surprise.

Between 20 and 40% of rural households regularly receive remittances.

Roughly estimated, we can assume that on average about 25% of the monetary income available in rural areas of sub-Saharan Africa comes from remittances and that between 20 and 40% of rural households regularly receive remittances.

6.1.1.1 New Technologies "Mobile Cash"

The rapid expansion of telecommunication technology in sub-Saharan Africa is well documented in research: According to Mothobi and Grzybowski (2017: 2) there were 600 million mobile phone subscribers in Africa in 2014, which represents a penetration rate of 68%. In the 11 sub-Saharan countries they researched, they documented a tremendous increase in mobile phone contracts, from 9.3 million in 2000 to 396.3 million in 2014 (Mothobi and Grzybowski 2017: 2).

Table 6.2 Remittances—quantitative clues from individual case studies

	Source	"Quantitative clues"
Africa	de Haan (2006: 7)	Remittances comprise approx. 25% of rural income
Sub-Saharan Africa	Godoy et al. (2012: 6)	32% of all adults in the eleven countries surveyed (about 80 million people) received remittances from internal migrants in the month prior to the study (4% of adults in the eleven countries, or about 10 million people, received international remittances in the month prior to the study) 20% of all adults in the eleven countries surveyed (about 50 million people) sent domestic remittances
Southern Africa	Reardon (1997: 739)	In areas near urban centers, remittances account for about 75% of nonagricultural rural income In areas far from urban centers, remittances account for about 20% of nonagricultural rural income
Burkina Faso	World Bank (2009: 16)	13% of households studied receive remittances from internal migrants
Ghana	Castaldo et al. (2012: 39)	82% of remittance payments in Ghana in 2005 and 2006 can be considered domestic remittances 53% of the remittance volume came from domestic remittances
	World Bank (2009: 12)	25% of households receive remittances from internal migrants
	Yaro et al. (2015: 121)	Almost 100% of the migrants working in the construction industry in Accra said that they sent money and goods to their families back home
Kenya	Godoy et al. (2012: 20)	Approx. 33% of the Kenyans surveyed traveled to bring money to friends or family members in the month prior to the study
	de Haan (2000: 17)	13–22% of the average income of urban-dwelling Kenyans is sent to their native home
	Ramisch (2014)	Up to 20% of the income of migrants in Western Kenya is sent as remittances to their region of origin
Namibia	Greiner (2008)	Case study from rural Northwestern Namibia: 87% of rural households receive remittances from internal migrants On average, these remittances account for 33% of the income of the rural households surveyed
Niger	Rain (1999: 155, 117)	Remittances make up 20% of rural household income in dry areas (northern Niger) and only 2–6% in less arid areas (southern Niger)

(continued)

Table 6.2 (continued)

	Source	"Quantitative clues"
		38.3% of migrants in Maradi bring money for the household when they visit their home village
Nigeria	McKay and Deshingkar (2014: 16)	36.1% of the population receive remittances from internal migrants (19.2% of the population receive international remittances)
Rwanda	Smit (2012: 93)	74.4% of migrant households in the rural region around Kigali receive remittances from urban areas
	McKay and Deshingkar (2014: 16)	21.7% of the population receives remittances from internal migrants
South Africa	McKay and Deshingkar (2014: 16)	6.8% of the total South African population receives domestic remittances (0.1% of the population receives international remittances)
	Steinbrink (2007, 2009)	In a village in the Eastern Cape Province (former Transkei), 60% of the households regularly receive payments from internal migrants On average, these remittances account for 23% of the available monetary income of rural households 70% of migrant household members who earn an income regularly financially support their relatives
	Steinbrink (2007, 2009)	In an informal settlement in Cape Town, 80% of those migrants with an income stated that they send money more or less regularly to their home village (Eastern Cape Province) On average, they transfer 20% of their earned income to their region of origin
Uganda	McKay and Deshingkar (2014: 16)	26% of the population receives remittances from internal migrants (5.3% of the population receives international remittances)
	McKay and Deshingkar (2014: 19)	63% of remittances from urban centers in Uganda go to rural households 58% of remittances from rural areas of Uganda go to rural households

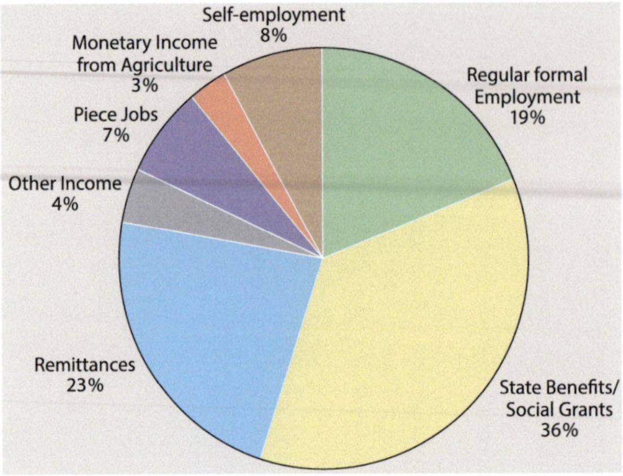

Fig. 6.2 Breakdown by income type of total monetary income of the resident population of a rural community in Eastern Cape province, South Africa. *Source* Modified from Steinbrink (2009: 238)

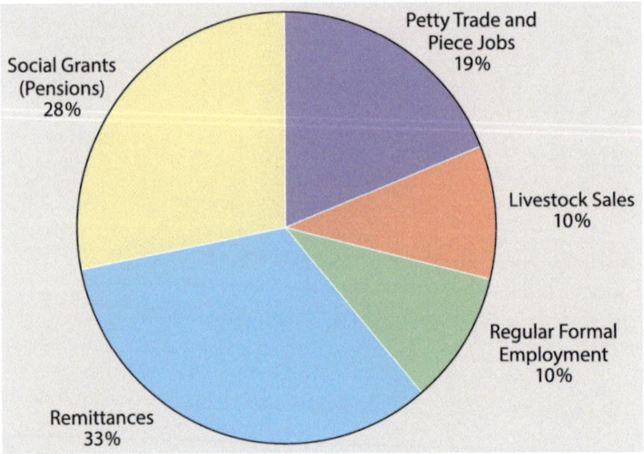

Fig. 6.3 Breakdown by income type of total monetary income of a rural community in Northwestern Namibia. *Source* Modified from Greiner (2008: 242)

This enormous growth in mobile telecommunication technology in sub-Saharan Africa has a direct impact on both translocal communication and the intensity of remittances made within translocal contexts, as well as the form such transfers take.[7]

---

[7]"At the same time, new information and communication technologies (ICTs) such as mobile phones, the internet and mobile money have facilitated new connections and decreased the physical and

Fig. 6.4 Poster by Airtel money advertising the easy (urban–rural) transfer of money via mobile phone. *Source* http://www.oceanogilvy.com/fr/images

For a long time, the system of remittances was shaped by a weakly developed network of bank branches and difficulties accessing banking services in most regions of Africa. Especially in rural areas, scarcely anyone had a bank account (Watkins and Quattri 2014: 7).[8] This is one reason why remittances are still frequently transferred in the form of cash today. In countless cases, the migrants themselves have brought money along during their (ir)regular visits or else entrusted it to fellow-migrant friends visiting the common home village. In some cases, bus drivers offer this courier service (see e.g., Steinbrink 2009; UN 2018: 137).

Both banks and mobile telecommunication providers have recognized the growing market for remittance transfer services and are increasingly establishing branches in rural areas and expanding their *mobile cash* systems, respectively. Among the *mobile cash* service providers are "*Airtel Money*" (see Fig. 6.4) in Burkina Faso and "*M-PESA*" in Kenya (Scharwatt and Williamson 2015). And people in Africa are making ever greater use of these practical options for *mobile money transfer*.

psychological distance between the city and the countryside." (Steel et al. 2017: 148)

[8]In addition, the fees for sending remittances via traditional service providers like Western Union or Moneygram are particularly high in sub-Saharan Africa compared to the world average (9.5% vs. 7.4% per US$200 sent, respectively) (KNOMAD 2016: iii).

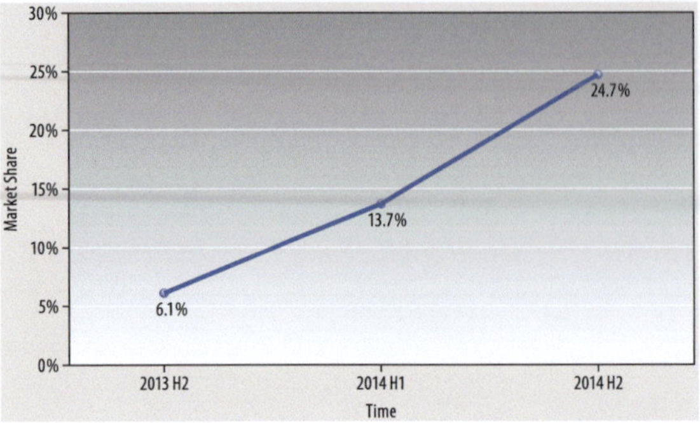

Fig. 6.5 Market share of orange money in formal remittances between Côte d'Ivoire, Mali and Senegal (2013–2014). *Source* Modified from Scharwatt and Williamson (2015: 13)

In Côte d'Ivoire (population: approx. 22 million), presently the largest market in West Africa for this form of transfer technology, the number of registered "mobile money accounts" had already reached 6.2 million by 2013. In Mali, too, the major provider "*Orange Money*" has experienced enormous growth in customer numbers. According to Scharwatt and Williamson (2015: 9), these "*mobile money*" transactions now account for about 20% of the West African state's gross domestic product. Figure 6.5 shows the increase in Orange Money's share of the formal remittance transfers between Côte d'Ivoire, Mali and Senegal within 1 year.

Founded in 2007, the M-PESA network in Kenya is now one of the world's largest providers of mobile transaction services. By 2014, 17 million Kenyans—or about two-thirds of all adults—had registered with M-PESA (Watkins and Quattri 2014: 17; Mothobi and Grzybowski 2017: 3). Recent figures (Mims 2013) show that, today in Kenya, 31% of GDP is spent via mobile phones (see Ramisch 2014: 33).

Inhabitants of rural areas especially benefit from "mobile cash".

There are indications that this new transfer technology is particularly benefiting rural areas in sub-Saharan Africa. Scharwatt and Williamson (2015) findings confirm this for Burkina Faso:

> The service has had particularly strong traction in rural Burkina Faso, where 60% of recipients live. MTN and Airtel expected their service to be popular in urban areas where remittance flows are well documented, but they had not anticipated such success in rural areas, where formal money transfer channels have a limited footprint and people prefer informal options to send money. Interestingly, rural

> customers tend to receive money more regularly than their urban counterparts. (Scharwatt and Williamson 2015: 20)

Some authors believe that the rapid proliferation of mobile phone technology in Africa has increased the actual volume of remittances within translocal contexts. As Youngstedt (2013) writes:

> According to World Bank (2009) data, Nigerien workers sent home remittances totaling US$14 million in 2000, US$66 million in 2005, and US$78 million in 2007 and 2008 (it might not be coincidental that these increases parallel the accelerated adoption of cell phones). (Youngstedt 2013: 150)

It is plausible to assume that the sum of transferred capital should increase as both the organizational and technical aspects of remitting funds are simplified. This supposition is supported by studies which show that frequent direct mobile phone contact with their rural relatives considerably increases the social pressure for migrants to regularly transfer funds (see Ramisch 2014). In this respect, we can reasonably expect the remittance rate to increase as the mobile communications market grows. To this extent, the inhabitants of rural areas of origin stand to benefit from this technological development.

6.1.1.2 Who Benefits?

Remittances can have positive economic effects—both direct and indirect—on migrants' regions of origin.

In academic literature, it is virtually undisputed that remittances can have positive economic effects on migrants' regions of origin: First, payments can contribute directly to improving the standard of living of the beneficiaries; second, greater purchasing power can strengthen endogenous economic cycles in the receiving regions. Supporters of the latter point tend either to emphasize the fact that the transfers enable recipient households to invest additional capital in agricultural or nonagricultural activities, or they cite the oft-touted "trickle-down effect." According to this hypothesis, the increased consumption of those families that receive remittances also benefits the other families, as it increases the demand for labor (production and services) (Ajanovic 2014: 104). For this reason, Tacoli (2004: 2) assesses the effects of the remittances on rural development and rural poverty reduction decidedly positively:

> Overall, synergy between agricultural production and urban-based enterprises is often key to the development of

> more vibrant local economies and, on a wider level, to less unequal and more ›pro-poor‹ regional economic growth.

Labor migration and remittances can amplify disparities and socioeconomic stratification.

However, other researchers call attention to the fact that labor migration and remittances can amplify both disparities and socioeconomic stratification. Thus, the question is: Who benefits from remittances and in what ways?

First, it is plausible to assume that the most vulnerable groups do not employ the strategy of labor migration, as migration often entails additional expenditures for the sending household and is thus associated with financial risk (see Ambrosius et al. 2008: 3). Therefore, in view of their risk sensitivity and existential security concerns, those most vulnerable households must forego dispatching labor migrants and cannot benefit directly from remittances.

Among those households that can afford to send migrants, relative wealth correlates with relative success: The labor market chances of better-trained jobseekers also tend to be greater in sub-Saharan Africa, and the probability that these migrants can regularly send part of their income (or larger sums) generated in the area of arrival to their rural relatives increases accordingly.[9] Since, in Africa too, levels of education correlate very strongly with the economic status of a household, remittances can actually exacerbate existing wealth disparities in the regions of origin (see Sect. 5.3.4.4).[10]

> […] wealthier households are more likely to benefit from migration since they are more able to educate their children and to send them over longer distances in search of work, and their children are generally more willing or able to remit. Migration is a riskier strategy for poorer households who are more likely to suffer in the absence of productive household members. In consequence, the effect of rural to urban migration and urban to

[9]"Survey results among migrants in Windhoek and Walvis Bay clearly support this: migrants with better education have higher incomes and send more remittances." (Greiner 2011: 615)

[10]"[…] the volume of the remittances, however, depends strongly on professional position and income: those who transfer large sums to the farm work as real estate agents, managers, technicians, teachers, accountants and nurses. Those who send less or have no support relations at all to the farm often have part-time jobs in retail or work as maids and security guards. The income of those who send transfer payments to the farm is on average 25% higher than that of those who do not send anything to the farm, and the amount of the transfer payments increases significantly with the monthly household income of the sender. (Pearsons r: 555**, N = 33)." (Greiner 2008: 232–232) [translated].

> rural remittances is likely to increase rural inequality. (Waddington and Sabates-Wheeler 2003: 9–10)

Regarding the significance and volume of remittances, various studies explicitly address the differences between wealthier and poorer rural households. From his study on pastoralists in Northwestern Namibia, Greiner (2008) arrives at the interesting conclusion that remittances constitute a greater share of resource-poor households' total income:

> The proportion of remittances within the overall household budget is thus significantly higher in households with a below-average income than in households with an above-average income. (Greiner 2008: 241) [translated]

Still, the absolute value of remittances tends to increase as income rises:

> The higher the households' income, the greater the sums of money that family members transfer from the cities. (Greiner 2008: 242) [translated]

Remittances constitute a large portion of income for poorer households but lose importance as income increases.

These findings align with Evans and Pirzada's (1995: 70) assessment that remittances mainly account for a large share of poorer households' income and lose relative importance as income increases. As Peil and Sada (1984) demonstrated more than 30 years ago, the volume of remittances rises in direct proportion to the incomes of migrant-sending households, thus exacerbating inequalities. Other authors who have focused on this topic[11] come to similar conclusions, so we may well consider these findings to represent a general pattern.[12]

By incorporating the influence of a migration networks' maturity, De Haas (2010) expands this point of view: He agrees that so-called pioneer migrants (or expanders, see Sect. 5.3.3) do generally come from more affluent households, meaning that the positive impact of transfers is initially felt only by those households that are already better off. But, as the translocal network grows and matures, the costs and risks of migration fall, leading to a decrease in migrant selectivity, thus mitigating the early, disparity-amplifying effects of remittances (de Haas 2010: 249).

---

[11]While Steinbrink (2009) finds no significant differences in the proportional share of remittances in the total income of different rural wealth groups in South Africa, he confirms that the absolute value of remittances is significantly higher for higher income households (Steinbrink 2009: 243).

[12]See, among others, Rain (1999: 180f).

Labor migration aimed at economic diversification is a strategy employed more by poorer households.

In sum, the research on this question indicates that it is not chiefly wealthy rural households that pursue economic diversification through labor migration and that benefit from remittances as a result. Poorer (though not the poorest) households also employ the strategy of translocal economic diversification. However, their strategic action is motivated less by the logic of benefit maximization than by a rationality of risk minimization (see Sect. 4.2). For that reason, it is important to clearly differentiate between these wealth groups.

To examine the extent to which remittances have a lasting positive impact on the economic development of rural regions and whether growth impulses and trickle-down effects can be expected, it is helpful to focus on the use of the transferred capital.

6.1.1.3 The Use of Remittances

Consumption or investment purposes of remittances

With regard to the economic consequences of remittances, it makes sense to distinguish between the *purposes* of consumption and investment. Are the transferred funds being used directly to cover needs at the rural household locations? Or are investments being made in the region of origin that serve to increase the prosperity of the households, ultimately promoting local or regional economic growth?

Most of the studies we have examined come to the conclusion that the majority of remittances are not invested in "productive assets" but used directly to secure or improve recipients' standard of living (see e.g., for Kenya: Ekbom et al. 2001; Olson et al. 2004; Greiner and Sakdapolrak 2012[13]; for Namibia: Greiner 2008, 2011[14]; for South Africa: Steinbrink 2009; for Niger: Rain 1999[15]; for Senegal:

[13]"[…] members from resource-poor households are more likely to migrate, but their remittances are often small and used for immediate consumption needs." (Greiner and Sakdapolrak 2012: n. p.)

[14]"Cash remittances are used to cover basic needs and are consumed immediately. There is hardly ever any money left to invest in livestock or home improvement, let alone better schooling." (Greiner 2011: 615–616)

[15]"Roughly half of the migrants reported sending cash gifts back to the village, though the size of these gifts is sometimes quite small. Much of this remittance money is used to purchase food and other necessities." (Rain 1999: 152)

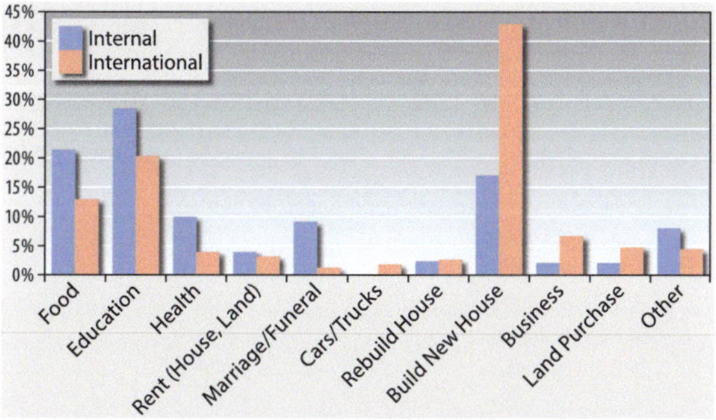

Fig. 6.6 Use of remittances in Uganda. *Source* Modified from McKay and Deshingkar (2014: 20)

Mortimore and Tiffen 2004[16]; for Ethiopia and Zimbabwe: UN 2018: 138; Litchfield et al. 2018). Only a small proportion of the transferred funds are saved or invested in an entrepreneurial sense (e.g., Ambrosius et al. 2008: 3–4).[17]

Different uses of the remittances of international and internal migrants.

For Uganda, McKay and Deshingkar (2014: 20) provide a more precise breakdown of the use of remittances (see Fig. 6.6). The findings on spending behavior shown in Fig. 6.6 confirm that a large proportion of the money transferred is used for basic needs such as nutrition and housing, as well as for education and health. The figure also illustrates some clear differences in the use of international and domestic remittances, which result from the fact that international migrants tend to come from better-off households. Because such households are less likely to rely on remittances to cover basic everyday needs (e.g., food), they can spend transferred money on the construction of additional or

[16]"[…] remittances mainly fund consumption needs, housing, and festivities." (Mortimore and Tiffen 2004: 271)

[17]In her work on migration and gender in Ghana, Awumbila (2015: 137) reveals that the patterns of spending remittances can vary depending on the gender of the sender and the recipient; she cites numerous other studies to underline this: "Furthermore studies indicate that in many societies, the gender of the person who sends and receives remittances can influence the use of that cash remitted. Remittances sent by women tend to be used for the immediate needs of households, such as food, health and education, while remittances of men are used for long-term investment in productive businesses and property (Quisumbing 2003; Pickbourn 2011; Lopez-Ekra et al. 2011). In Ghana, Pickbourn's (2011: 152) study in the Savelugu-Nanton District of Ghana, showed that households in which the main remitter or recipient of remittances were women tended to spend more on education of their children than other migrant households."

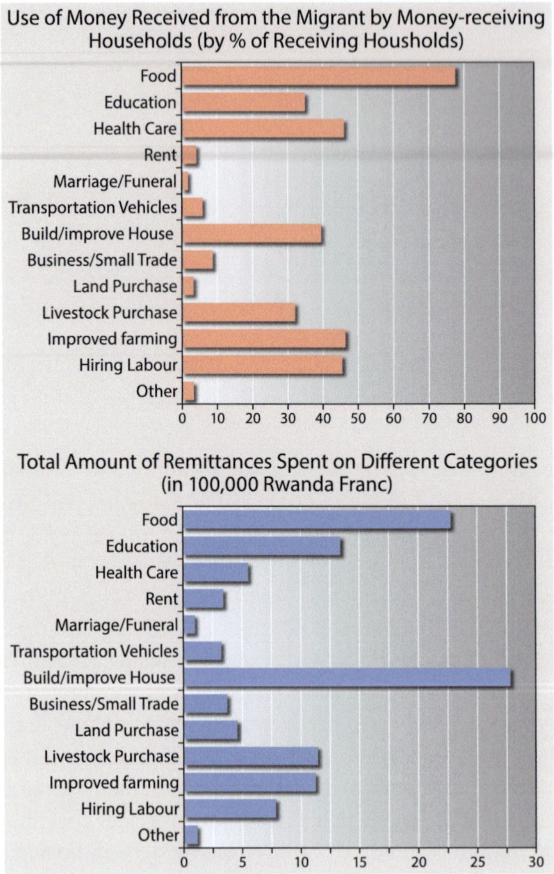

Fig. 6.7 Use of remittances in rural migrant households in Rwanda. *Source* Modified from Smit (2012: 100)

better housing, for example. Strikingly, however, the share of international remittances used for productive investments is also relatively low.

In his study on rural migrant households in Rwanda, Smit (2012: 100) reports slightly different results (see Fig. 6.7):

Here, too, we see most of the remitted funds channeled into the categories "build/improve house", "food", "education" and "healthcare".[18] At the same time, however, the figures

[18]"40.2% of the money-receiving households use the remittances to pay health insurance, indicating a general lack of financial resources within rural households to pay the 'mutuelle'." (Smit 2012: 63)

indicate some considerable investment in economic activities, especially in the agricultural sector ("land purchase", "livestock purchase", "improved farming", "hiring labor"). This also explains Smit's (2012: 79) observation that better-off households' agricultural production benefits more from migration than that of poorer smallholder households.[19]

For Kenya, the oft-cited research by Tiffen et al. (1994) shows that remittances from rural–urban migrants can play a central part in the process of agricultural intensification (see Sect. 6.1.3).

For West Africa, Oltmer (2015: 15) summarizes that most migrant remittances are spent on food and health care for the family, as well as on school fees for children, who would be unable to attend school without these funds. Traditional celebrations such as weddings and funerals are also financed with funds earned elsewhere. In addition, remittances are used for the purchase or construction of the real estate. Oltmer (2015) offers an example of this from a remote region east of the Nakambé (Burkina Faso) whose housing structure was characterized by traditional straw huts until a few years ago; now, numerous brick houses have been built or financed by migrants.[20]

---

[19]"[...] better-off households experience significantly more often an increase in agricultural production, as a result of migration, than poorer households. So, 70% of the poorest households indicated to have experienced a decrease (or a 'strong decrease') in agricultural production after migration, in contrast with 16.2% of households from better-off household group [...]. The explanation for this relationship again can be found in the fact that better-off households have higher household income and get more often money from the migrant household member, which can be used to hire laborers and to buy seeds, unnatural fertilizers or livestock to produce natural fertilizers." (Smit 2012: 79)

[20]This region, which is mainly inhabited by members of the Bissa ethnic group, has a well-developed migration network closely tied the tomato harvest in Italy. When migrants return home every year between August and December, large amounts of money flow into the region. Because of the economic upswing caused by this capital influx, the area is also known as the "Côte d'Azur" of Burkina Faso. Living conditions here are on average much better than in the rest of the country; however, inflation is also higher, as there is relatively little production in the region but a great deal of money. This has led to an increase in prices and thus in the cost of living. Consequently, those families who do not have migrant members living in Italy or elsewhere are at a significant economic disadvantage (Oltmer 2015: 14).

Ultimately, there is no unequivocal analytical distinction between consumer and investment spending of remittances at the household level.

The fact that remittances are mainly used for consumption sometimes leads authors to conclude that these payments do not have a positive impact on economic development. However, it should be noted that there is ultimately no clear analytical distinction between consumption and investment spending at the household level. Expenditures for education, better nutrition, and health can have a positive effect on the (future) workforce and thus on the economic opportunities of the household. Thus, such expenditures can be regarded as an investment in the future, even if they formally fall under consumer spending.[21] Furthermore, even when used entirely for the purposes of consumption, remittances increase the overall demand in the region of origin, thereby inducing—in the best-case scenario—investments in the entrepreneurial sector (Ambrosius et al. 2008: 3–4). However, to reap such benefits, it is essential that the products consumed are locally or regionally produced and that the services demanded are provided locally; otherwise capital flows out of the region without stimulating endogenous economic cycles.

Moreover, we must bear in mind that remittances are also part of the total income of migrant families, so their use cannot generally be distinguished from other sources of income. Presumably, the transferred money is not used in a fundamentally different way than other available funds. Instead, we can assume that the spending behavior for remittances reflects the usual spending behavior of households, which, in turn, depends strongly on the respective economic situation:

Migrant households with a very low total income must spend a large share of it on satisfying basic needs; only when these needs are met are funds available for investment purposes.

Migrant households with a very low total income must spend an especially large part of it on satisfying basic needs; only when these needs are met are funds available for investment purposes. Thus, if—as illustrated above—the funds generated at the migrants' destinations are primarily used for consumption to satisfy the basic needs of the rural household members, we can reasonably surmise that most rural households are using their remittances to close systemic "subsistence gaps" or to meet the need for resources in emergency situations (Dorlöchter-Sulser 2014: 43). This indicates that very few households use the strategy of translocal economic diversification (see Sect. 4.2) with the aim of maximizing

[21]So too argues Awumbila (2015: 137): "However it can be argued that investment in food, education and health are important for alleviating poverty and thus for enhancing development."

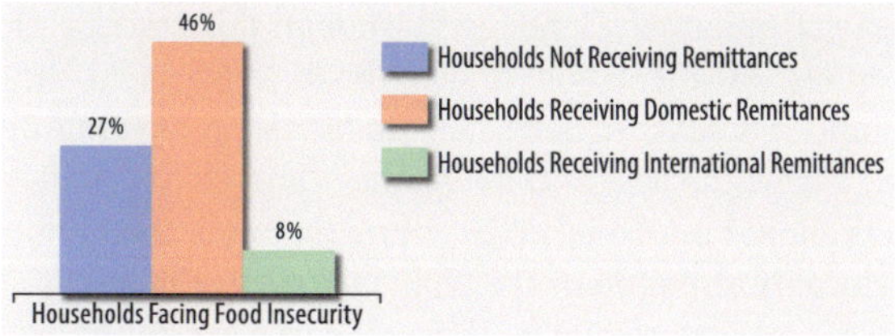

Fig. 6.8 Remittances and food security. *Source* Modified from World Bank (2009: 18)

benefits in order to accumulate capital for future investments in income-generating activities.

That the translocalization of livelihoods and remittances do not necessarily lead to a higher standard of living is revealed in the results of a World Bank study (see World Bank 2009: 45). Based on the *Ghana Living Standard Measurement Survey* (2005).[22] this study shows that households without remittances generally have better quality housing than those receiving remittances from domestic migrants; the degree of electrification of houses is also higher among non-translocal households. In terms of educational attainment, too, this study shows that translocal households with remitting internal migrants are on average less educated than locally organized households. Only in households receiving remittances from international migrants are the prosperity indicators more positive than in locally organized households—particularly if the migrants are residing in an OECD country.

A study from Ethiopia reveals that translocally organized households receiving remittances from internal migrants are evidently more vulnerable to food crises than locally organized households (46% to 27%; see Fig. 6.8).[23]

Translocalization is above all a strategy of vulnerable households.

These findings are no indication that translocal economic diversification leads to a reduction in the standard of living; rather, they show that vulnerable households tend to employ

[22]See IHSN Survey Catalog (2005).

[23]Here too—as in the study from Ghana—it is evident that households receiving international remittances are significantly less vulnerable to food crises. It can be assumed that these households could already be counted among wealthier population groups before the migration of the remitting household members.

translocalization as a strategy.[24] Though this strategy clearly represents an adaptation of livelihood systems to the vulnerability context, it does not, under the prevailing conditions, enable the actors to overcome vulnerability. Still, these results should not be taken as contradiction of the notion that translocality strengthens the resilience of livelihood systems, since the results provide no indication of what the situation of those households might look like without migration and remittances.

Considering the conclusions of the available studies, we must assess the general impact of migration and remittances on prosperity and poverty reduction in rural areas as rather small. Translocally diversified livelihood systems, based on the combination of subsistence production, market production and wage labor, are mostly oriented toward the objective of risk minimization and thus might reduce vulnerability. However, for a great number of the translocal households in sub-Saharan Africa, the financial resources available to them are not enough for sustainable investment (or else the households regard such investments as too risky). For that reason, remittances are mainly used for consumption, rarely stretching beyond the satisfaction of immediate and basic needs of the rural population.

This can also answer the question of why many small farmers who now live in translocally diversified livelihood contexts are in too poor a position to quickly and effectively mobilize existing potentials for agricultural intensification (see Sect. 6.1.3).

No "economic miracles" to be expected: Remittances do not stimulate strong endogenous growth dynamics in local rural economies.

At present, the capital flow of remittances cannot be assumed to stimulate sustainable endogenous growth dynamics in local rural economies. Productive investments in the agricultural or nonagricultural sector are only being made by a very small number of households, most of whom were economically better off even without remittances. Only investments by this group could be expected to stimulate

[24] "There is no evidence of a rural poverty trap (Barrett et al. 2001). The extreme and very poor households are instead well represented in rural-urban migration flows because they feel forced to seek opportunities elsewhere driven by the structural constraints in their livelihoods. The proposed entry barrier is resolved with loans or selling of assets, consequently these migrants are ill-prepared for high living costs in urban areas." (Schutten 2012: 116)

growth[25] (see, among others, Olson et al. 2004). As long as a large share of rural households are operating at the subsistence level under conditions of high vulnerability, remittances are unlikely to bring about an "economic miracle" in rural sub-Saharan Africa.

6.1.2 Workforce and Workload

One effect of the translocalization of livelihoods is the altered demographic structure of residential populations in rural areas, which is characterized by a disproportionate presence of women, children and the elderly, as described in Sect. 5.3.4. This demographic imbalance appears when much of the able-bodied population—particularly young and innovative people—leave rural areas, hoping for more favorable employment opportunities elsewhere. Various studies describe the resulting population structure as a demographic anomaly that inhibits growth and significantly hampers economic development in rural areas.

Given the persistence of labor-intensive farming practices in most rural areas of Africa, migration leads to a shortage of family labor, which may entail a reduction in farming capacity.

> The vast majority of all migrant households experienced increased workload as a result of the migration of household member(s). This is surprising because in areas with high population density [...], the outmigration of people is expected to result in relief in terms of less underemployment. However, more than 50% of all respondents from migrant households indicated to have 'higher workload' after migration and 6.2% even said to experience 'much more workload' [...]. (Smit 2012: 77)

The rural-urban labor division mostly results in increasing agricultural tasks for women besides their usual household tasks. (Smit 2012: 79)

Often women take over land cultivation—a process described as the "*feminization of agriculture*".[26] Consequently, women regularly perform double duties in

[25]"Wealthier households are also highly engaged in diversifying their income through off-farm activities – more often in situ than by migration – and invest in the farm by buying additional land, labor and other agricultural inputs." (Greiner and Sakdapolrak 2012: n. p.)

[26]See, among others, Lastarria-Cornhiel (2008), Ajani and Igbokwe (2011), Abdulkadir and Abdullahi (2018), as well as Blasche and Inhetveen (1983).

terms of production and reproduction: They look after children, the sick, and the elderly, cultivate the fields and take care of the livestock (see De Haas and van Rooij 2010; Smit 2012).[27] But the labor of the elderly and children is also employed to compensate for the loss of agricultural workforce.

Considering the case of Burkina Faso, Oltmer (2015) describes the impact of the migration of the able-bodied on the distribution of workload in rural areas. Migration results in an extremely low proportion of men in the population of some regions of Burkina Faso[28]; in turn, the share of old people and children there is disproportionately high. Accordingly, migration decimates the number of family members capable of working, i.e., the "*bras valides*", the strong workforce for fieldwork. In response, children often leave school (periodically or permanently) to devote their labor to agriculture in order to shore up the family's often precarious food situation (Oltmer 2015: 13–14).

Agricultural productivity can be weakened by translocalization.

In addition to the negative consequences of this process on education, the rural workforce shortage also makes it difficult to intensify smallholder agriculture. And, as a rule, migrant households usually lack the economic means to compensate for the shortage of labor through improved inputs or cultivation methods.[29] Only a few better-off migrant households manage to hire laborers for agricultural or domestic tasks (De Haas and van Rooij 2010: 55).

"Lack of agricultural initiative" as a result of translocalization?

Moreover, the selective pattern of who migrates and who remains behind can result in a loss of agricultural knowledge and the slow adoption of new agricultural knowledge and innovations (see Sect. 6.1.3).

Some studies report that, in the context of the livelihood translocalization, a certain "lack agricultural of initiative"

---

[27]In a study on Niger (Niamey), Youngstedt (2013: 136) writes about the tasks of the wives of male migrants: "Effectively serving as household heads to ensure social reproduction, many take on difficult burdens of extra work when their husbands fail to send remittances, while others assume responsibility for investing remittances." For Rwanda, Smit (2012: 26) describes the situation as follows: "The departure of especially young and physically powerful man, often results in an increasing workload for staying women in the household. Besides the household duties and the care for the children, they namely have to work in agriculture and to solve all arising household problems."

[28]In some cases, the gender ratio there is 7:10.

[29]The research by Tiffen et al. (1994) shows, however, that remittances have played an important role in the process of agricultural intensification in Kenya.

develops among the rural population, which in turn leads to a reduction in productivity and thus a greater economic dependence on nonagricultural income earned elsewhere. In an interview, an expert on rural development in Burkina Faso described this phenomenon as follows:

> The system of remittances often creates an asymmetry, because the money simply comes and people no longer take the initiative themselves. They just wait and don't work anymore. They just sit there waiting for the money to come. (Personal interview with a member of the Ministry of Economics and Finances Burkina Faso, 2015 [note: author's own translation from French])[30]

Studies of Western Kenya (Siaya) and of migrants from Cape Verde in Italy describe intrafamilial tensions: Migrants are subject to immense social pressure to send money on a regular basis, while those who remain at home become increasingly dependent and less willing (or able) to innovate. Those who remain behind may even neglect farming, leading to food shortages, which, in turn, increases the pressure on the migrants (ACP 2012: 11; Greiner and Sakdapolrak 2012).

However, other studies suggest that remittances can sometimes compensate for the loss of the workforce by providing for the hiring of external laborers to work the land (see e.g., Cotula and Toulmin 2004; Greiner 2008; Smit 2012).[31] With regard to Embu (Kenya), Greiner and Sakdapolrak (2012) demonstrate that more than half of translocal households employ nonfamily agricultural workers and conclude that "Remittances enable households to compensate loss of labor" (Greiner and Sakdapolrak 2012: n. p.).

Smit (2012) points out, however, that in many cases the remittances simply do not suffice to offset the lack of workforce.[32] In other words, this strategy only benefits economically better-off households with successful migrants—i.e., only those families, who because they already enjoy the necessary economic leeway, can afford economic

[30]The interview was conducted by the author H. Niedenführ in Burkina Faso.

[31]For example, Cotula and Toulmin (2004) indicate that many rural households in Senegal hire cheap labor from central Mali to compensate for the workforce shortages that arise from the labor migration of young men to neighboring countries and to France.

[32]See, among others, Vargas-Lundius et al. (2008).

diversification (whether local or translocal) in pursuit of benefit maximization anyhow.

In summary, in Africa South of the Sahara, translocal livelihoods often lead to a shortage of family labor in agriculture, which remittances rarely offset. This may lead to a reduction in the capacity for work and innovation that many smallholder producers would need in order to respond adequately and flexibly to the demand incentives of the (global) market. Viewed in this light, the translocal organization of livelihoods hinders the effective use of the most important resource in rural areas—the land. Not only do market potentials remain untapped, but food crises and soil degradation are becoming increasingly probable. Since wealthier households can more easily compensate for workforce shortages (e.g., through remittance-financed investments in technology, external labor or improved inputs) and thus exploit market advantages, the widespread shortage of agricultural labor likely exacerbates socioeconomic disparities in rural areas.

6.1.3 Innovation and Knowledge

Translocality means networks and mobility between places, which encompass not only the spatial movement of people, goods and capital but also the *mobility of knowledge* (see Sect. 2.3). It is along network lines that translocal communication—and with it, the flow of knowledge—takes place, and it is along translocal network corridors that knowledge-carriers migrate, transferring their knowledge from one local context to another.

The mobility of ideas, values, and knowledge in translocal networks has far-reaching effects in many areas.

The mobility of ideas, values, and knowledge within translocal contexts influences such diverse areas as education, health, agricultural production, political participation, and gender relations[33]—thus directly affecting the living conditions of households.[34] The economic effects of translocal knowledge transfers can be observed both at the level of the household and, in some cases, the macroeconomic level. As such, these effects have the potential to trigger major changes and developments (see Calì and Cantore 2010). The interplay

[33]See Sect. 6.3.

[34]However, immaterial remittances are under-researched in comparison to financial or in-kind remittances (De Haas and van Rooij 2010: 44).

between these "social remittances" (Levitt 1998; Levitt and Lamba-Nieves 2011) and translocal livelihoods is clearly reflected in the area of knowledge transfer and innovation:

Translocality can (a) accelerate the acquisition and dissemination of new knowledge in transmigrants' rural areas of origin (e.g., knowledge about [agricultural] technologies and practices), while at the same time (b) disrupting or even destroying the existing knowledge base (e.g., knowledge about agriculture).

Opportunity to increase agricultural productivity through knowledge transfer.

(*On a*) Transmigrants often acquire valuable knowledge at their place of arrival, learning new modes of action and sometimes acquiring knowledge about new technologies. They transfer this knowledge to their areas of origin, whose local economies it can benefit. Various studies show that this translocal transfer of knowledge can increase agricultural productivity, among other things. Oltmer (2015: 14) describes an exemplary case from Burkina Faso: There have long been strong migratory movements from the region east of the Nakambé to Italy for the tomato harvest. Early on, the harvesters from the region used the know-how they acquired in Italy to introduce the "Italian" technologies in their home region, thus optimizing tomato cultivation there. Today, the increase in productivity in this market segment is of great benefit to the rural regional economy. Such examples show that knowledge transfer through translocal networks can contribute to agricultural productivity and increased yields.

Other studies indicate that it is not only the agricultural sector that benefits from migrants' experiences. Knowledge acquired in the course of migration, can—as Hahn (2004) points out in his study in Burkina Faso—be as simple as basic know-how about handling money, an essential foundation for a variety of economic activities in rural households:

> More important from the perspective of those left behind is the option of reintegrating their relatives with these experiences into local society. For the people in the village community it is a fundamental advantage if their sons or brothers know other ways of life. This other way of life includes, for example, dealing with money, which the migrant worker is truly trained for. (Hahn 2004: 395–396) [translated]

Deshingkar (2004) stresses that the knowledge acquired in migration can be useful for establishing nonagricultural enterprises. Returning migrants may bring back new skills, ideas, and ambitions that can help them to set up nonfarm businesses (Deshingkar 2004: 13). In a case study from

Somalia, Watkins and Quattri (2014) found that 80% of all newly established business ventures are funded by remittances and based on the ideas of migrants (Watkins and Quattri 2014: 15).

Migration can lead to an "outflow" of knowledge.

(*On b*) Translocality entails not only the influx of new, potentially economically useful knowledge into rural areas; migration in translocal contexts can also lead to an "outflow" of knowledge. This is particularly true in the agricultural sector: The translocal organization of households can lead to the loss of agricultural know-how that has been handed down from generation to generation. A kind of "culture of migration" often develops in translocal contexts, in which jobseeking in the city is an integral part of the gender roles of young men (and increasingly, young women too). In such contexts, the danger is that time-tested agro-ecological knowledge will no longer be passed on from the elders to young people, because the latter initially consider this knowledge irrelevant.[35] But when they then return to the countryside (e.g., at retirement or during periods of prolonged urban unemployment), they lack the necessary agricultural know-how and agro-ecological knowledge on soil conditions, precipitation patterns, etc. Ramisch (2014) provides an excellent example of this in a case study from Kenya entitled "We will not farm like our fathers did", where he quotes a migrant who has returned to the countryside:

> My father and grandfather worked this land together and could tell you anything about it. Every bird, the way the dew forms in the morning, the tunneling of worms – everything had its story. When I was little I wasn't interested to learn any of that and so I tried my luck in town. I have nothing like that to share with my sons, even if they wanted to learn from me. Now that I am retired, I try farming but I see that the rain does not fall when the old men say that it should and our soil is tired and sick. We don't know the land and we don't know the rains the way we used to. (Ramisch 2014: 28)

The loss of agricultural knowledge reduces the rural population's flexibility in response to opportunities and/or threats (e.g., on the agricultural market).

The loss of agricultural knowledge reduces the rural population's flexibility in response to opportunities and/or threats (e.g., on the agricultural market) (see, Misiko 2007; Shisanya and Khayesi 2007; Doss et al. 2006):

> […] migrants may indeed grow less knowledgeable about the environments they have left behind (or only infrequently visit),

[35]"The ones who move away are not interested in learning the old ways [of adaptation or coping], neither are they contributing new ideas that might save us today." (Quote from an older farmer; in: Ramisch 2015: n. p.)

> and are similarly less aware of or able to track changes in those environments. Similarly, household members only temporarily resident (or only told by phone of the most pressing news from the farm) are less likely to be able to effectively judge the emerging opportunities and constraints for farming [...] or to weigh them against other, non-agricultural livelihood pursuits [...] in a given locale. (Ramisch 2015: n. p.)

The loss of traditional agro-ecological and agricultural knowledge not only means that existing opportunities for intensifying or expanding production are not effectively used; it can also lead to a decline in productivity (even under normal climatic conditions). In turn, households' coping capacities in times of crisis are reduced, decreasing their resilience (e.g., to food crises). The result: Their vulnerability increases, as does their dependence on nonagricultural sources of income (and further labor migration).

The relationship between migration, translocality, and knowledge flow is thus ambivalent. Further comparative research in different regional contexts is necessary for a more accurate estimation and appraisal of the economic advantages and disadvantages of knowledge in- and outflows in translocal contexts.

6.2 Ecological Dimensions of Translocal Livelihoods

To date, the ecological dimension of translocal livelihoods has received too little scientific attention. The focus here lies on the relationship between translocality and the changing natural environment. So far, this nexus has been treated rather one-sidedly, namely as part of the discussion around environmental and climate-induced migration: How do environmental changes affect migration? In this debate, a distinction can be made between "alarmists" and "skeptics"[36] (see

[36] Jónsson (2010), for example, takes a "skeptical" view in his meta-analysis: "From the review of case studies on environment and migration in the Sahel, it appears that environmental stressors such as drought do not necessarily lead to migration. This is usually because migration – particularly long-distance and international migration – requires resources and during drought, resources are scarce." (Jónsson 2010: 11). This also explains the tendency towards internal rather than international migration observed at such times. Moreover, Jónsson

Greiner et al. 2014: 5; Pott and Felgentreff 2016: 73). While the former attempt—in a neo-geodeterministic fashion—to establish a direct causal relationship between environmental change and migration, viewing "climate migration"[37] as an unavoidable "emergency response" (see e.g., Bogardi and Warner 2009), the skeptics do not at all see a direct causal link and reproach the "shaky empirical character and sloppy nature"[38] of alarmists' arguments (Piguet 2013: 155).

Several recent studies on environmental and climate-induced migration argue that vulnerability, adaptation and resilience are closely linked to the livelihoods, risk exposure and adaptability of individuals, households, or groups (Adger 2006; Folke 2006; Gallopín 2006; Sward 2016). In these studies, migration is seen either as a failed adaptation to environmental change, as an attempt by individuals and/or households to reduce their vulnerability to environmental stress, or as an important component of livelihood security (McLeman and Smit 2006; Tacoli 2011; Scheffran et al. 2011; Warner and Afifi 2014).[39]

Today there is an increasing consensus that environmental changes do not have a determining effect on migration processes but that different, complex, and interlinked forms of social inequality contribute to vulnerability to environmental

Environmental change has no determining effect on migration processes. Instead, different, complex and interlinked forms of social inequality contribute to vulnerability to environmental change—and thus to migration.

(2010) points out that reactions to environmental changes are by no means universal, nor clearly predictable: For Mali, for example, Findley (1994) describes an increase in the "short distance" migration of women and children in times of drought, whereas for Niger, Afifi (2009) find that women tend to stay behind while men tend to migrate. Carr (2005) writes for Ghana that younger men tend to migrate, whereas older men try to fulfill their role as head of the household for as long as possible (Jónsson 2010: 13).

[37]On the problematic concepts of "environmental migrant" and "climate refugee" in the debate on climate change, see Aufenvenne and Felgentreff (2013) as well as Aufenvenne et al. (2013).

[38]For a literature review, see Morrissey (2011) as well as Obokata et al. (2014).

[39]See, for example, the research projects "TransRe—Building resilience through translocality. Climate change, migration and social resilience of rural communities in Thailand", "Where the rain falls" and "micle—Migration, climate changeClimate change, and environment. Social–ecological conditions of population movements in Mali and Senegal".

change—and thus to migration.[40] These intersecting factors include, for example, age, ethnicity, education level, occupation, and gender (see Hummel 2015). With regard to the complex nexus of climate change, land degradation, and migration in certain regions in Mali and Senegal, the *micle* project has shown, among other things, that schooling is one of several important factors that influence migration decisions in the face of unstable socio-ecological conditions.[41] Overall, the project's studies show that the fewer financial means and less social capital people have to offset income losses stemming from climate-related crop losses or failures, the greater their vulnerability to environmental changes is (e.g., rising precipitation variability). For some vulnerable groups, migration can indeed be the only way to ensure survival (see van der Land and Hummel 2013).

The causes of "climate-induced migration" are not primarily climatic, but structural!

In fact, the studies show that "climate-induced migration" is not primarily climate-induced[42]: It is largely dependent on the specific capabilities and adaptation opportunities of individuals, households, and groups (see e.g., Schade 2013) —which are determined far more by structural than by climatic factors. Migration is always a social phenomenon and not an environmental one!

If one considers migration as a process rather than a state, a reciprocal relationship becomes clear: Ecological conditions affect migratory movements, and conversely, migration shapes environmental conditions. So far, however, the *effects* of migration and translocal livelihoods on the natural environment in rural areas of sub-Saharan Africa have not been sufficiently investigated. What feedback effects do different

[40]"Predicting the impact of climate change on population distribution and movement is fraught with difficulties. However, it seems unlikely that the alarmist predictions of hundreds of millions of environmental refugees will translate into reality. What is more likely is that the current trends of high mobility, linked to income diversification, will continue and intensify. Past experiences suggest that short-distance and short-term movements will probably increase, with the very poor and vulnerable in many cases unable to move. Underlying these trends is the growing need for the diversifi cation of income sources, and the spatially unequal distribution of economic opportunities." (Tacoli 2009: 523)

[41]"The lower the formal level of education, the higher the probability of economic activity in smallholder agriculture." (Hummel 2015: n. p.)

[42]"Environmental changes and hazards represent an inherent part of the local context, but are only one factor influencing migrant trajectories." (Ziegelmayer and Spaan 2019: 52)

aspects of translocality have on soil quality and other ecological parameters?

A holistic consideration of the mutual relationship between environmental changes, agricultural land usage and translocality is necessary. Synthesizing the concept "*socio-ecological systems*" with a translocal perspective, as proposed by Greiner et al. (2015), may provide a suitable approach to this end. In justifying their idea, the authors point out that researchers have long neglected the role migration plays in sustainable resource management. For too long, resource management in the areas of migrant origin has been regarded as part of a vicious circle of poverty, which stems from failed local adaptation ("in situ adaptation") negatively affecting the resource base (e.g., O'Keefe 1983); hence, migration has been interpreted as a threat to the environment and the agrarian base (e.g.,Warner et al. 2010, Myers 2002).[43] According to Greiner et al. (2015: 5–6), a rethinking of this process has only recently begun with the emergence of *(social) resilience* approaches.[44] Resilience studies have aimed to take greater account of the complexities of the interplay in human–environmental interactions. One focus, for example, has involved investigating the influence of social networks on adaptation processes in natural resource management (see Bodin et al. 2005; Pelling and High 2005; Rodima-Taylor et al. 2012, as well as Tompkins and Adger 2004).

With their idea of combining the *socio-ecological systems* approach with the concept of translocality ("*translocal socio-ecological systems*"), Greiner et al. (2015) have sought to enable researchers not only to consider the effects of environmental change on migration (and translocalization) but also to analyze the feedback effects of migration (and translocalization) on the environment. In addition, this approach explicitly spotlights actors' practices in dealing

---

[43]See Greiner et al. (2015: 5–6).

[44]"During recent decades, resilience and related concepts, such as adaptive cycles, multistable states, panarchy, nested scales, and response diversity, have become popular to describe complex SES [Socio-Ecological Systems]. These concepts urge us to understand the complexity and the dynamics of human–environment interactions in a more comprehensive way, and address linkages across scales explicitly as important elements for the functioning of SES. As such they generally also place more emphasis to the influence of social networks on the adaptive management of natural resources." (Greiner et al. 2015: 6)

with “environmental stress” across spatial distances (Greiner et al. 2015: 9–10).

On the question of whether migration and translocality have a positive or negative impact on the natural resource base in the areas of origin, the available empirical studies provide very different answers. This variance becomes clear when looking at studies on Kenya, for example (see Greiner and Sakdapolrak 2012).

For instance, Evans and Ngau (1991) present a study indicating a positive impact of translocal livelihood (and migration) on the environment (soil quality, etc.). They find that those rural households in the Kiriyanga District that have a translocally diversified income portfolio are also more productive in farming: Because of the remittances they receive, they are better able to take financial risks, e.g., by investing in more intensive or diversified cultivation. Olson et al. (2004) also give a positive assessment. They show that successful commercial farming based on remittances ultimately correlates with a greater willingness to invest in soil and water protection measures.

Further, Tiffen et al. (1994), as well as Mortimore and Tiffen (2004), show how additional income (e.g., from remittances) offers farmers in Machakos the chance to reinvest in agriculture, enabling them to expand their terrace cultivation for example. They point out that this has contributed to a recovery of the environment in previously degraded areas.[45] Lewis (1985) indicated a similar situation for the Central Highlands Region in the 1980s, pointing out that the least erosion damage was found around those farms involved in *cash crop* production (e.g., coffee) with higher investments in soil and water protection, made possible by migrants’ remittances.

Contrary to these positive assessments, there are numerous studies that stress the negative impact of a translocal livelihood on farming and natural resources. For example, Ekbom et al. (2001) report that the highh migration rates among men from the Murang’a District force those left behind to take on ever more agricultural tasks.[46] Their higher workload results

---

[45]Greiner et al. (2014: 12–13; based on Tiffen et al. 1994).

[46]Lohnert (2002) and Steinbrink (2009) describe precisely this phenomenon for the South African context in their respective studies on the multilocal and translocal household structures linking the rural Eastern Cape Province and townships in Cape Town. They show that the rural population hardly cultivates the land available to them, not least because

in the neglect of labor-intensive water and soil conservation measures. Conelly (1994) comes to a similar conclusion on Rusinga Island; despite an increasing population, the emigration of young male workers has prompted a deintensification of agriculture, leading to the abandonment of terraced fields and the discontinuation of other erosion control measures, which has in some cases brought about complete crop failures. Regarding the impact of remittances on soil protection measures in a comparative study of Embu, Meru, and Machakos, Nyangena (2006) finds that investments in erosion control and soil conservation decrease when households receive remittances. He presumes the fact that rural household members are also reorienting themselves in the course of translocalization to be a central reason: Agricultural activities are accorded less importance within the economically diversified household portfolio; thus questions of soil quality and resource protection are also losing relevance among the rural residential population (see Nyangena 2006: 3:21).[47]

The very divergent empirical results regarding the of migration on the environmental conditions in rural areas of origin indicate a need for more research in this area. At the same time, a review of the available studies conveys the suitability of the translocal livelihood perspective (see Chap. 3), as it is a research framework that approaches migration as an integral component of livelihood security and as a strategy for adaptation. As Sakdapolrak et al. (2016) point out, such an approach also offers an opportunity to analytically align livelihood-related migration research more closely with research on *socio-ecological resilience*[48]:

of the shortage of workforce. Often, however, the uncultivated land is not made available for other farmers' use but is instead monitored by specially employed guards, rendering it a totally idle resource (see Schmidt-Kallert and Kreibich, n.d.). The fields lie fallow, and inadequate pasture management and too frequent grassland burnings result in severe erosion damage, reduced biodiversity, and severe soil degradation.

[47]"Households with Remittances are less likely to adopt soil conservation measures as indicated by the negative and significant coefficient. A possible explanation is that the extra earning opportunities reduce the time for farm work or relax liquidity constraints […]. Additionally, they may have little concern about land quality due to their orientation towards off-farm activities." (Nyangena 2006: 3:21)

[48]For comparison, see Folke et al. (2010) and Oliver-Smith (2012).

> [...] the link between social and ecological spheres remains poorly addressed. While in most conceptualizations environment is considered as a threat, a comprehensive understanding needs to address both sides – environment as a threat as well as a source of benefits for society. Despite referring to the concept of resilience, literature on migration as adaptation has developed in remarkable isolation to current research on socio-ecological resilience [...]. This is a missed opportunity [...] since these concepts would provide a fruitful basis for operationalizing complex social and ecological interactions beyond one-sided notions of ecosystems as either 'threat' or 'resource,' while at the same time shedding light on linkages across scales and societal processes of adaptation and transformation. (Sakdapolrak et al. 2016: 86)

"Translocal social resilience"*: migration, vulnerability, and resilience in the context of environmental change*

Responding to the call of Black et al. (2011: 449) for a more comprehensive understanding of the significance of migration to vulnerability and resilience in the face of environmental change, Sakdapolrak et al. (2016) propose the concept of *translocal social resilience* as a framework for analysis; such a concept is hardly conceivable without a notion of translocal livelihoods. The authors from the project *TransRe* outline the advantages of the concept of *translocal social resilience* as follows:

> Conceptualizing *translocal social resilience* in this way emphasizes i) the everyday practices of social actors who are embedded in social fields, which are structured by the endowment of social actors with different forms of unequally distributed capital and at the same time are ii) embedded in translocal social networks that facilitate the flow of resources, practiced and ideas between places. An approach to translocal social resilience as constituted by practices in translocal social fields therefore places emphasis on the circumstances under which connectedness emerges, is maintained or nonexistent. Furthermore, a translocal social resilience approach to the environment-migration nexus stresses iii) agency of mobile and immobile actors with regard to the articulation, regulation and distribution of ecosystem services and hence unravels the constructed and politicized nature of human-environment relations. (Sakdapolrak et al. 2016: 89)

Ambivalent diagnoses: Some studies show that migration has destructive effects on natural resources and socio-ecological systems, while others come to the opposite conclusion.

In summary, there is currently no general consensus on the impact of migration (and remittances) on land use and thus on natural resources in rural sub-Saharan Africa. Some studies indicate that migration has destructive effects on the resource base and socio-ecological systems, while others come to the opposite conclusion. Despite—or perhaps because of—the contradictory empirical results of individual studies, the assessment prevails that the effects are not only very complex but above all context-dependent. This means two things: First, there is increasing acknowledgement that people's action differ according to their respective vulnerability situation and that the *outcomes* of their strategies (including their migratory actions) therefore differ too.[49] On the other hand, the realization is gradually gaining ground that contexts of action are increasingly formed within social fields that transcend spatial boundaries. While there is still a tendency in research to interpret environmental migration as a one-time unidirectional "*emergency response*", there is an increasing number of studies that explicitly acknowledge the embeddedness of migrants and nonmigrants in translocal social and socio-ecological networks (Greiner et al. 2014: 7–8).[50]

Thus, it becomes apparent that a translocal livelihood perspective that addresses the context-variance of action orientation (see Sect. 4.2) can also help resilience research to overcome its container-spatial bias and analyze the context-dependence of socio-ecological consequences of the action. Future research on the relationship between migration and environmental change must therefore examine resilience to environmental stress against the backdrop of translocal realities.

6.3 Social Dimensions of Translocal Livelihoods

Translocality and shifting values regarding education, health, gender, etc.

Numerous studies on migration in Africa touch on the various facets of how translocal livelihoods affect social practices and processes. These include the (re-)negotiation of gender roles and shifts in social norms, values, and attitudes as well as approaches to education and health issues. In a study on

[49]Migrations based on a lack of alternatives are very likely to have different consequences than migrations with the objective of benefit maximization. Though the achievement of the latter's goals is by no means guaranteed, failure in the first case is presumably much more devastating.

[50]"[…] migrant social networks can help to build social capital to increase the social resilience in the communities of origin and trigger innovations across regions by the transfer of knowledge, technology, remittances and other resources." (Scheffran, Marmer and Sow 2011: 1)

"*social remittances*" in Tanzania, Helgesson Sekei et al. (2014) find that

> [...] among emigrants, return migrants and the friends and relatives they left behind, attitudes and practices related to education, health, employment and business, gender, the environment, politics, human rights and the distribution of wealth can positively change, thanks to the social remittances transferred in the framework of a migration experience in the South. (IOM 2013: 49–50)

In the following, some of the areas mentioned will be considered in terms of the influence translocal livelihoods have on them. Our focus lies here on *education* (Sect. 6.3.1), *gender* (Sect. 6.3.2), and *health* (Sect. 6.3.3).

6.3.1 Education

Education is generally regarded as the key to social development. For this reason, the Millennium Development Goal 2 "*Primary education for all children*" was formulated, which aimed for all children to be able to attend primary school by 2015. It is estimated that about half of the world's children without access to education live in Africa. In 29 countries worldwide, the net school enrollment rate is below 85%—16 of them are in sub-Saharan Africa. In countries such as Burkina Faso, Cote d'Ivoire, and Mali, the enrollment percentage is less than 70. With an average net enrollment of 77%, sub-Saharan Africa has the lowest rate in the world, although significant progress has been made (for comparison, the net school enrollment rate in 1999 was 59%). Formally, the educational situation in Africa has improved; the United Nations even speak of "spectacular successes" in the African school sector: By 2011, 25 of all 53 African countries had achieved enrollment rates of at least 80%, and 11 more had reached at least 75%.[51] Between 1999 and 2011, the total number of children in Africa who do not attend school at all fell by almost half, from 40 million to 22 million. In general, school enrollment rates continue to rise, albeit less sharply in recent years. Still, only 62% of pupils finish primary school

---

[51]Eleven countries (including, for example, Burkina Faso, Liberia, Eritrea and Niger, as well as Nigeria, the most populous country) had school enrollment rates below 75%.

on average—it is a mere 28% in Chad and only 32% in Angola (see UNESCO 2012: 58f.).

Despite the progress made in recent years, the educational situation in Africa remains extremely inadequate. In addition to the low enrollment rates, the high number of early dropouts and the frequent shortfalls in educational attainment indicate the educational plight in large parts of Africa (see Langthaler and Scharer 2005: 9). The problems and challenges are manifold: the limited availability and poor quality of the teaching materials and textbooks; in some cases high school fees, the poor quality of school infrastructure; the sometimes-long distances to and from school; deficiencies in the training, payment, and motivation of teachers and frequent absences of teaching staff. All of these factors play just as important a role in the predicament as the children's health, family background, nutritional situation, and social or gender-specific disparities.

Especially in Africa's rural regions, there are immense deficits in education.

It is particularly in the rural areas of Africa that educational problems culminate: There, school enrollment rates are disproportionately low (in part due to the long distances between schools), school dropout rates are particularly high, and the inequality of opportunity between the sexes is most pronounced (Verspoor 2008: 16).

Education is an important migration-inducing factor.

The question then arises what influence the translocal organization of livelihoods has on the educational situation in rural sub-Saharan Africa. In the empirical case studies available, this aspect is rarely placed in the foreground; indeed, there is a clear dearth of research in this field.

The relationship between education and translocality has already been addressed in various chapters of this book. First, we stated that education, or rather school attendance, is an important migration-inducing factor: On the one hand, adults migrate so that their remittances can enable other members of the family (usually their own children or younger siblings) to attend school; on the other hand, children move in order to take advantage of better school infrastructure. This means that either the pupils themselves migrate in order to receive (better) schooling or their education is made possible or improved through funds generated via migration.

Studies addressing the use of migrant remittances show that the financing of education has a high priority in spending practices. Aklilu and Catley (2010), among others, report that one central use of remittances in the Borana and Somali regions of Ethiopia is to pay school fees for younger siblings.[52] The aim is to give children better opportunities on the labor market, so that in their turn, they can send money to open up better educational opportunities for future generations:

> In both Borana and Somali areas, education was regarded as the best means of pursuing an alternative livelihood. By education, informants meant access to paid employment. It was explained how employed family members could help by sending remittances home, and in helping younger siblings to attend school and become employed. A Sheikh in Shinile commented that a single employed family member is '... worth 30 head of camels that do not perish away in a drought.' (Aklilu and Catley 2010: 32)

Educational strategies as safety-oriented rational action

This quote clearly reflects the orientation of the translocal households' actions toward risk minimization: educational strategies as a safety-oriented rational action. According to the authors, this strategy can ideally set in motion a positive spiral of school attendance, educational success, labor market opportunities, income generation and investment (see Aklilu and Catley 2010: 32).

Many children and adolescents are on the move within translocal contexts to gain access to education.

For translocal households in Tanzania, Dilger (2013) shows that there is a clear expectation among rural household members that members working in the city invest part of their income in the education of their younger relatives. This can be done by financing school attendance in the rural area of origin through their remittances or by taking the children and adolescents in at the urban household location to give them access to education there (Dilger 2013: 117).[53] The latter case is especially likely in cases where schooling is not possible at all in rural areas or its quality is judged to be inferior. Multiple studies indicate that many children and adolescents are on the move within translocal contexts in sub-Saharan Africa in order to attend school away from their rural homes (see, among

---

[52]In a study of Ghana, the *Migrating out of Poverty Research Programme Consortium* (2016b: 3) arrives at a similar conclusion: 62% of the migrants interviewed in Accra stated that migration had improved the education level of their children and family members.

[53]"In particular, rural and urban families expect family members who work or trade in the city to support their parents, uncles, and siblings in times of need and crisis and to reinvest some of their acquired wealth in the education and economic advancement of their younger relatives [...]. This may involve rural families sending their children to their more wealthy relatives in urban areas who take care of their children for an extended period of time and provide for their schooling." (Dilger 2013: 117)

others, for Kenya: Beguy et al. 2010; for Namibia: Greiner 2008; for South Africa: Lohnert 2002 as well as Steinbrink 2009; for Tanzania: Dilger 2013; for Burkina Faso: Niedenführ 2018; for Ghana: Ziegelmayer and Spaan 2019). They are usually sent from their villages of origin to relatives or household members in urban areas, but educational migration also takes place from the city to the countryside.[54] There are various reasons for education-related transmigration:

(a) Families have the prospect, or hope, that relatives at the other locality will be able and willing to bear the costs of schooling and living expenses for schoolchildren when these costs exceed the financial capacity of the rural-dwelling household members.
(b) Schoolchildren are sent to urban areas, if no one living at the rural household location can adequately care for them, their education and their upbringing.
(c) The educational opportunities and the quality of schooling in the city are assessed to be better. This is particularly true if there is no (secondary) school in the regions of origin. The progression from primary to secondary school is often only possible through migration.
(d) Children are sent to cities for schooling because the schools there are often seen as more prestigious than those in the countryside and have a better reputation in terms of educational attainment (and thus success later in life).

Effects of "social remittances" on educational practices.

That migration (experience) affects educational practices is underscored by a large-scale study conducted by the International Organization for Migration (IOM). For Cameroon, for example, the study shows that families with returned migrants spend more on education than families without migrants, whereby the expenditures on education are highest among households with a member currently in migration (IOM 2013: 27, 30; see also Zourkaleini et al. 2013).[55] This result can be interpreted in two ways: either

[54]Beguy et al. (2010) show in a study of two slums in Nairobi that many children born there are sent to rural areas because the school conditions and opportunities in the villages are judged to be more advantageous. Lohnert (2002) describes a similar situation in South Africa. Ziegelmayer and Spaan (2019) also describe the urban to rural migration of children for educational purposes in Ghana.

[55]"Expenditure on […] education is also higher among households with return migrants than among non-migrant households. It is highest in households with members who are currently abroad." (IOM 2013: 30)

(a) that migrants are generating sufficient income to cover educational expenditures (school fees, school uniforms, learning materials and living expenses) via remittances or (b) that migrant families are experiencing a shift in values that leads to a greater appreciation for the benefits of formal schooling ("*social remittances*"). Still, it must be noted that the results of this IOM study refer to international migration.[56] As shown in Sect. 6.1.1.3, spending patterns vary significantly between households with international migrants and those with domestic migrants. The latter are economically weaker on average, and their rural household members must often use most of the remittances to cover their immediate subsistence needs. Nevertheless—as the studies listed in Sect. 6.1.1.3 show—households with domestic migrants also spend a considerable proportion of their remittances on education. This can certainly be considered a sign of appreciation for formal education.

The opportunities for advancement (elsewhere) associated with labor migration offer an important incentive for education.

It is plausible that translocality and the opportunities for advancement (elsewhere)—which the rural population often see in labor migration—offer a concrete incentive for education. The curricula and the benefits of formal school education often relate more to urban life and work settings than to rural and agricultural contexts. If migration is regarded as an opportunity to improve one's own living conditions or those of one's family, there is additional motivation to invest in education. In this respect, Batista et al. (2010) come to an interesting conclusion in the case of the Cape Verde Islands. The authors find that more highly qualified workers' better chances of migration lead to an accumulation of human capital: 40% of university graduates surveyed stated that they would not have enrolled at university had they not had the opportunity to migrate internationally (see also Calì and Cantore 2010).[57] This idea can be applied to the domestic context as well: In rural areas, households are often more willing to invest in (secondary) school education if migration

[56]"Households with a member living abroad spend more on education." (IOM 2013: 27)

[57]"Evidence on skilled migration from Cape Verde suggests that migration has encouraged the accumulation of human capital. Almost 40% of Cape Verdean university graduates would not have enrolled in university had they not had the opportunity to migrate [...]. Importantly the authors find that a shock decreasing migration by 9 per cent reduces the educational attainment of non-migrants by 7 per cent." (Calì and Cantore 2010: 9)

offers opportunities to capitalize on the knowledge (or degree) acquired. However, there are not (yet) any studies available on this subject.

Translocalization can also worsen the educational situation.

As described in Sect. 6.3.1, migration and translocal household arrangements can actually lead to a worsening of the educational situation in some cases, for example, when young adults or children are kept out of school in order to compensate for the migration-induced shortage in agricultural workforce or if they must contribute to securing the translocal livelihood through their own labor migration (see Sect. 6.1.2).[58]

Although the translocality of livelihoods tends to positively affect education by providing many children in Africa with access to (better) schooling, which they would otherwise be denied, it does not follow that translocal households automatically have a higher average level of education than locally organized ones. For Ghana, for example, a World Bank study (World Bank 2009) based on the *Ghana Living Standard Measurement Survey* (2005) shows that the average level of education in translocal households with remitting domestic migrants is actually lower than in locally organized households; only for households receiving remittances from international migrants can a positive correlation with educational indicators be observed.

This discrepancy manifests the clear and self-perpetuating connection between the economic situation and the education level of households on the one hand and the context dependency of the motives for translocalization on the other: More vulnerable or poorer households (with lower average levels of education) are translocalizing because the local opportunities available to them are few to none, and their lack of resources prohibits them from engaging in international migration (especially to OECD countries). Economically better-off households (with higher average levels of education), on the other hand, have the freedom to choose not to translocalize or to make strategic use of the opportunities that arise in other places—possibly also abroad—according to the maximization rationality. As Etzold (2017) points out, spatial

[58]This is confirmed by empirical studies from Ghana: 19% of men and 7% of women in Accra stated that their level of education would have been higher if they had not migrated to the city. They indicated that they had dropped out of school to move to Accra to find work and support their families (Migrating out of Poverty Research Programme Consortium 2016b: 2).

mobility may indeed go hand in hand with social mobility—but this is mainly true for better-off households (Etzold 2017: 46). Thanks to the remittances of economically successful migrants, wealthier households can spend more money on education and thus improve their chances on the labor market. Conversely, the most vulnerable households simply lack the necessary (financial) means for such mobility and are thus unable to engage in education migraton—which prevents their upward social mobility (Etzold 2017: 46).

Translocalization can contribute to educational inequality.

In this respect, while migration represents a strategy for adapting to the extremely precarious school situation in large parts of Africa, and translocal structures are often used to overcome the obstacles on the way to formal school education, translocality is not likely to remedy existing inequalities but rather to reinforce them.

6.3.2 Gender

There is an agreement in the literature *that* migration and the translocality of livelihoods affect existing gender relations in sub-Saharan Africa. However, regarding the question of *how* migration and translocality influence gender relations and the evolution of gender roles, the studies at hand reveal (sometimes widely) diverging perspectives.[59] Some authors argue that the role of women is strengthened by translocal ways of life (e.g.,Rocheleau and Steinberg 1995; Rocheleau 2001; Awumbila 2015); others see the consequences for women in a more critical light (e.g.,Ramisch 2015; Youngstedt 2013; De Haas and van Rooij 2010; Francis 2002).[60]

The "feminization of migration" and the "feminiziation of agriculture".

In the discussion on translocality/migration and gender in sub-Saharan Africa, two main strands can be identified: On the one hand, the migration of women is addressed (often under the catchword "*feminization of migration*"); on the other hand, the focus lies on the shifting roles and tasks of women left

[59]"The discussion indicates that migration can be both a source and an effect of greater empowerment of women especially in urban areas and in areas of origin. However despite this potentially positive impact, internal migration and in particular rural-urban migration has been viewed generally negatively by policy makers because of the widely held perceptions that it can lead to negative outcomes for migrants, their families, their areas of origin or destination." (Awumbila 2015: 141)

[60]See also Greiner and Sakdapolrak (2012).

behind in rural areas, following the labor migration of male household members ("feminization of agriculture") (see Sect. 6.1.2). The central questions in this debate are therefore:

1. To what extent does the *migration of women* affect (strengthen or weaken) their familial or social position?
2. To what extent does the *migration of men* affect (strengthen or weaken) the position of the women who remain behind?

6.3.2.1 Women's Migration in Translocal Contexts and Its Influence on Gender Relations

The "feminization of Migration".

In Sect. 5.3.4.1, we already discussed the fact that the migration systems in Africa, long dominated by male labor migration, have recently undergone significant changes and become more diversified in the course of translocalization (see also Sect. 5.3.3). One of the most striking changes is that ever more women in Africa are involved in migration. This process is thus often referred to as the "feminization of migration".

As for the significance of this process for the social role of women, two fundamental lines of questioning arise:

1. Is the migration of women evoking a change in the prevailing gender relations in sub-Saharan Africa? Or is the increased migration of women in sub-Saharan Africa itself an expression of already shifting values? Or is the feminization of migration perhaps best understood as a manifestation of a mutually reinforcing and advancing process of reciprocal influence between shifting values and migration?
2. Might the "feminization of migration" ultimately have much less to do with value shift and the liberation of women from the corset of traditional gender roles than development researchers would like it to? Is female migration perhaps primarily an expression of vulnerability, which results not only from the social position of women but also from the vulnerable situation of their whole households (including female *and* male members)?

Female migration and gender-related shifts in values.

Ultimately, the question arises whether, in the course of the feminization of migration, existing gender relations are changing in a way that gives women more opportunities for self-determination and social participation, or whether women are taking their roles and "gender-specific" patterns of action with them when they migrate. Do women migrate to achieve more independence (social, economic, cultural, etc.), or do they migrate to meet the expectations of their existing roles? How does female migration affect the (re-)production and transformation of gender arrangements? By no means do the case studies at hand provide unanimous answers.

Emancipatory effect of female migration.

Various studies emphasize the *emancipatory effect of female migration*. This argument can be found in numerous older works, too. Pittin (1984); Gugler and Ludwar-Ene (1995) and Ouédraogo (1995) regard the emancipatory flight from gender-specific social obligations and the liberation from rural traditions as partial causes of female migration. With regard to Ghana and Tanzania, Cottyn and Scharpendonk (2013) argue similarly, presenting migration as an opportunity for women to free themselves from the "straitjacket" of traditional gender relations. The increasing mobility of women in sub-Saharan Africa is thus interpreted as an opportunity to change and redefine conventional gender roles:

> Conventional gender roles may gradually change as a result of the increasing autonomy of women who break out of their traditional gender straitjackets. (Cottyn and Scharpendonk 2013: 3)

Werthmann (2007: 316) too sees migration as a way for women to gain more autonomy. Using the example of young mothers from rural areas of Burkina Faso who move to the mining regions as migrant workers, leaving their children behind with the grandparents, she describes how work gives these women the opportunity to earn the "*dot*" (dowry) needed to "buy their freedom" out of unwanted marriages—or to meet a wealthy man who will assume this cost for them.

The studies that highlight the emancipatory effects of female migration mostly refer to cases in which women migrate alone, i.e., not within existing intra-household networks. According to Gugler and Ludwar-Enes (1995: 261), independent female migration in Africa has been neglected by scholars for too long[61]; in migration studies, women have primarily been perceived as the wives of circular migrant workers, who either remain behind or follow their migrant husbands.

[61]"The independent rural-urban migration of women has been grossly neglected in African studies to date." (Gugler and Ludwar-Ene 1995: 261)

In their instructive literature overview, Gugler and Ludwar-Ene (1995) distinguish three types of women migrating alone in the rural–urban context:

(1) *Young, unmarried women with little formal education.* They often find their first job as domestic workers in the city and usually live with a man with whom they are in a relationship. This form of migration is often circular. Guigou and Leriollais (1992) show for the Serer region in Senegal that in the 1950s unmarried girls and young women were already migrating to Dakar in the dry season in order to earn money there as housemaids. When they reached marriageable age, however, they returned to the countryside.[62] Hamer (1981), Grosz-Ngaté (1991) as well as Anker and Hein (1986) report similar patterns in Mali and Nigeria.
(2) With the increasing expansion of secondary schooling outside of larger cities, ever more *better-educated young women* are moving to cities to find jobs and capitalize on their education there (see Hollos 1991, Wurster 1995). Many of them are trying to postpone or avoid marriage and are striving to improve their individual standard of living (at times, with the help of male partners). Corresponding case studies are available for Accra and Lomé (see Seibert 1995; Dinan 1983).
(3) *Widowed, divorced, or separated women* are the third group of African women who move alone to urban areas.[63] For Ethiopia, Baker (1994) shows that although women have equal land rights, agricultural activities are effectively an exclusively male domain. Cultural, physical, and financial barriers prevent women from plowing with oxen, as is the custom there.[64] Divorce rates in Ethiopia are high;

[62]"In Senegal, for instance, young Serere and Diola women have sufficient autonomy to migrate seasonally to the city to work as domestics, until they marry, whereas Toucouleur, Peul and Soninke women have no such independence to leave the village." (Brockerhoff and Eu 1993: 561)

[63]"The share of separated, divorced or widowed migrants is higher among women migrants compared with men, except for in Burkina Faso, and almost one quarter (23 per cent) of women migrants from Zimbabwe are separated, divorced or widowed." (UN 2018: 70)

[64]This situation in Ethiopia is unusual as in large parts of Africa hoe cultivation is still the dominant production method in smallscale farming, and women play a very important role in fieldwork (see, for example, SOFA Team, Doss 2011). Abdulkadir and Abdullahi (2018: 5395, 5397) show for Kano, Nigeria, women are involved in all kinds of

and given the limited nonagricultural income opportunities in rural areas, cities offer better—if not the only—alternative livelihood sources for divorced or separated women. In Bamako/Mali in the 1980s, 11% of female rural–urban migrants were divorced women or widows. In the case of Kampala (Uganda) too, it has been reported that marital problems are the main motive for women migrating on their own (other important reasons: divorce and husband's death). Many women in Nigeria are also said to flee from unwanted marriages to the cities; there they work as "*karuwai*" (courtesans) or independent traders. Older widows are sometimes driven to urban areas by the fear of being suspected of witchcraft (see Ludwar-Ene 1986).[65]

It is questionable whether these three types of rural–urban migrants can actually be described as "independent" and if their migration can be understood as an expression of emancipation. Only the second type ("*young, better-educated women*") suggests such an interpretation. Migrations of the first type ("*unmarried girls and women with little formal education*"), on the other hand, are presumably best understood within the context of vulnerable households' translocal livelihood security strategies. That through their migration these girls and young women can break free from fixed role structures and develop greater autonomy, therefore, does not seem immediately plausible; after all, they often remain entangled in dependency relationships in the city as well as in (translocal) patriarchal household structures. Migrations of the third type ("*widowed, divorced and separated women*") can hardly be interpreted as an expression of freedom of action; rather, they appear to represent a form of flight, a largely indispensable course of action in situations of

farming activities "after clearing and ploughing by the men," although they rarely possess land. The authors found that only 36.1% of women own land, 60.3% of whom are widows, 44.4% divorced women, 25% single women, and 23.3% married women. The authors explain these figures with the patrilineal inheritance of land, the practice of married women putting their land "under the custody of their husbands" and the financial capital constraints that prevent women from buying their own land (Abdulkadir and Abdullahi 2018: 5397–5398).

[65]See also Offiong (1982) for South-East Nigeria.

personal crisis.[66] Whether leaving the region of origin has an emancipatory effect on these women ultimately depends on each woman's circumstances in her new surroundings.

Independent migration is much more difficult for women.

In our opinion, any overestimation of the direct connection between the "*feminization of migration*" and shifting gender relations in studies on sub-Saharan African is due to researchers overestimating both the degree of independence among women migrating alone and their quantitative share in female migration overall. Due to widespread vulnerability in many African life contexts, any form of migration is nearly impossible without social contacts. An "independent" migration is even more difficult for women, as they are structurally disadvantaged by their social position in many African societies. In addition, women have greater difficulties on the urban labor market and receive significantly lower wages (see Adepoju 2008 and Smit 2012). For this reason, embeddedness in social networks is likely even more essential to women than to men. Moreover, the "independent" rural-urban migration of women is often viewed with suspicion in rural areas, where "urban life" is often associated with decaying values and "immoral" behavior.[67] Consequently, for women, migration also carries the risk of losing one's good reputation and status. According to Schutten (2012),[68] in Rwanda (in the region around Kigali), it is often very difficult or even impossible for many women to return to their village of origin, because the urban labor market often confines poorly educated migrant women to the

[66] "The rapidly increasing rural exodus of women is only partly the result of emancipation, an opportunity to shape one's own life, rather it is primarily an expression of the rapid impoverishment of a section of the population that has to bear the brunt of social change." [trans.] (Vorlaufer 1985: 143)

[67] "Stiff competition and a certain amount of job discrimination in receiving countries have narrowed the range of job choices for migrants. Many find themselves in easy-entry jobs: most of the already disadvantaged women, given their lower educational background, have been driven into the 'sex trade'." (Anarfi 1993: 17)

[68] "In-depth interviews revealed that it is much more difficult for young woman to return to the rural household because to survive hard living conditions in the city some woman sold themselves and even if this was not true out of shame for this prejudice female migrants preferred not to return to the rural area unsuccessfully." (Schutten 2012: 113)

stigmatized work of prostitution,[69] and female migrants must reckon with massive scorn in their native villages, even if they do not personally engage in this activity at all.

Migrating women are less likely to leave the social fabric.

There are many indications that the increase in female migration, which has been observed for decades, mainly relates to the increasing importance of translocal livelihoods and their maturation. A large part of women's migration in Africa today likely takes place within translocal social fields and between the different locations of translocal households. In other words, female migration does not involve leaving a social structure behind (i.e., escaping or liberating themselves); it is much more a matter of spatial mobility *within* an existing social context, which spans across space. Thus, most women do not migrate to escape their role in the household; they migrate to *fulfill* their role within the *translocal* household.

Most women do not migrate to escape their role in the household, but to fulfill it within the translocal household.

However, the role of women in translocal contexts seems to be changing considerably, as their responsibilities are no longer limited to caring for children, the elderly and the sick (in rural or urban areas). Instead, women are increasingly involved in gainful employment. Today, significantly more women are migrating within translocal contexts to generate income (for their families) in urban areas:

> Women migrants are increasingly drawn to the wage labour market (both formal and informal) as a survival strategy to augment meagre family income. (Adepoju 2005: 2)

Adepoju (2005) interprets this development less as an expression of a changed set of values than as a reaction to worsening poverty: "[…] female migration has become a major survival strategy in response to deepening poverty […]." The classical gender division of labor is also undergoing radical change, particularly because urban employment conditions for men are worsening, forcing women to look for additional income opportunities to provide for their families financially.[70]

[69]Steinbrink (2009) also mentions that male rural–urban migrants in South Africa often stress that they prefer "rural women" as spouses because "town girls" tend towards immoral behavior and are not familiar with "rural traditions".

[70]"The gendered division of family labour has also been upset by the loss of male employment through urban job retrenchment and structural adjustment, forcing women to seek additional income-generating activities to support the family." (Adepoju 2006: 33–34)

Increased autonomy of women through the "feminization of migration".

One essential factor in the growing importance of female labor migration as a livelihood strategy for translocal households is the rising level of education among women (see Sect. 6.3.1). It is only through formal schooling that young women have certain employment opportunities on the urban job market. In this respect, the schooling of girls living in rural areas is becoming a higher priority for households.

Although in most cases, female labor migration is more likely to reflect economic constraints than female self-determination, we can assume that the changing role of women within translocally organized livelihood systems will lead to a change in women's social positioning in the long run. That is because female labor migration clearly entails an increase in independence, for one because working women have their own income. In cases where they earn more than the male members of the household, this can be tantamount to a substantial shift in gender relations.[71] In Ghana, Senegal, and Nigeria, a growing number of cases have been reported in which married women migrate to the city to earn a monetary income for the household, leaving their children with their husbands.[72] Adepoju (2005) calls this development a "*turn-around in traditional sex roles*". However, Hollos (1991) argues for Nigeria that such a change in women's status is not primarily the result of migration per se; the decisive factors are education and occupation (see also Trager 1995).

More economic independence and thus greater autonomy for women evidently has a strong influence on patterns of remigration: As early as the 1980s, studies from Sierra

[71]On this point, see Awumbila (2015: 138): "For example they cite AbdulKorah's (2011) study among the Dagaabas in northern Ghana, which demonstrates how the sending of remittances by modern Dagaaba women is changing perceptions about traditional gender roles and power relations. Furthermore, Wong's (2006) study among Ghanaian women in Canada also shows that remittances enabled them to negotiate power relations within transnational households. However, Awumbila et al. (2015) argue that there are cases where the absence of husbands and receipt of remittances by women does not necessarily increase women's decision-making powers and may instead, produce conflicts between women and men over the management and use of remittances."

[72]"Anecdotal evidence reveals a striking increase in the numbers of women – who traditionally remained at home – leaving their spouses behind with the children, who, in a reversal of parental responsibilities, are looked after by their fathers or by other female members of the family. The remittances these women send home are a lifeline for family sustenance." (Adepoju 2008: 24)

Leone, Ghana, Nigeria, and the former Zaire showed that *female* rural–urban migrants tended to communicate a desire to live permanently in their respective cities (see the overview of Peil and Sada 1984). Research in Nigeria in the early 1990s also showed that more women than men favored spending retirement in the city over returning to the countryside.[73] With regard to age-specific rural–urban migration rates in urban areas of Ghana, Singlemann (1993) found that men over the age of 45 more often migrated out of the cities, while for women of the same age group migration into the city was still relatively pronounced. Likewise, for the rural areas of the Somali region in Ethiopia, Devereux (2006) has illustrated that the quantitative gender imbalance among older age groups tends strongly toward the male population, which he attributes mainly to gender-related differences in remigration patterns.[74] In a study on the migration patterns of inhabitants of two slums in Nairobi, Beguy et al. (2010) also conclude that significantly more older men return to their areas of origin than older women.[75]

Life in the city is often more attractive for women.

Gender differences in patterns of return migration suggest that women particularly seem to find urban life more attractive than rural life. For example, in his study in South Africa, Steinbrink (2009) reports that many rural–urban migrant women are glad not to have to do "typical rural women's work" like carrying water and collecting firewood (anymore).[76] Considering that women in rural sub-Saharan Africa spend a total of 40 billion hours a year fetching water—a figure roughly eight times greater than the comparable figure for men (FAO 2003)—it is not surprising that women appreciate life in the city.[77]

---

[73]While 36% of women would prefer to spend their old age in the city, this figure was only 20% for men (see Gugler and Ludwar-Ene 1995: 264).

[74]"The most skewed sex ratios occur in the older age groups. Among people over 60 years of age, the sex ratio peaks at 203, meaning that there are twice as many men over 60 as women." (Devereux 2006: 120)

[75]"After age 60, males are clearly more likely to out-migrate [out of the slums] than their female counterparts, reflecting gender difference in retirement age effect on migration." (Beguy et al. 2010: 564)

[76]Malawian women, for example, spend more than 9 h a week procuring water and firewood for their households (FAO 2003).

[77]Especially younger women emphasize that there is "simply more going on" in the city and that rural life is boring.

Yet, these results do not necessarily mean that women are more likely than men to break off their translocal connections. Rather, these findings merely show that, in some cases, women prefer to perform their tasks at the urban site of the translocal household, and that their growing economic independence in the course of increasing integration into the (urban) labor market enables them to take advantage of this opportunity.

"[…] women are more urban than men."

At present, it does not appear that the "feminization of migration" leads to the abandonment of translocal livelihoods in favor of a more sedentary lifestyle. Too great is the socioeconomic need for translocality which arises from the conditions of vulnerability in sub-Saharan Africa's rural *and* urban areas. Many households currently have no alternatives. However, should the living situation and the labor market conditions in urban areas improve substantially, the research results suggest that women would probably be inclined to give up the rural household base. In a nutshell, Gugler and Ludwar-Ene (1995: 263) summarize: "[…] women are more urban than men."

6.3.2.2 Women's Remaining at the Place of Origin and Its Influence on Gender Relations

Today, agricultural production, and food security in Africa lie predominantly in the hands of women: Over 90% of staple foods and over 30% of food crops are now produced by women, and in most regions, 50–80% of the agricultural workforce is female (Schäfer 2002: 1; FAO 2011: 5, 32). One cause of this is the translocality of livelihoods.

The influence of male circular labor migration on the gender-specific division of labor in rural sub-Saharan Africa is relatively well researched.[78] The focus lies on the observation that, in view of the unchanged labor-intensive farming methods prevalent in most rural regions of Africa, the migration of able-bodied men has led to a shortage of family labor at rural household locations (see Sect. 6.1.2). In order to make up for this shortage of labor, women are, therefore, often forced to take on agricultural tasks, which traditionally fall within the male sphere of responsibility according to classical

[78]See, among others, Jyrkiäinen (2014).

perceptions of gender roles. Today, many women are responsible for the entire production process. This phenomenon is often referred to as the "*feminization of agriculture*".[79]

Higher workload for women with the "feminization of agriculture".

To the questions of how the reorganization of task assignments affects the gender order and whether this might lead to a shift in values and a strengthening of the social position of women, the available studies offer no clear answer.

Numerous studies draw particular attention to the additional burden associated with the doubling up of productive and reproductive duties in the course of the "feminization of agriculture"[80]: Women continue to perform their "traditional" domestic tasks (e.g., caring for children,[81] the sick and the elderly), in addition to farming the fields and looking after the livestock (see, among others, De Haas and van Rooij 2010; Smit 2012). Schäfer (2002: 36) shows that the wives of low-income migrant workers as well as widows and divorced women are particularly burdened by the simultaneous demands of many overlapping areas of work. In addition, their difficulties in accessing resources (land, investment capital, services) and their limited control over the resource they do have reduce these women's chances of becoming more economically autonomous by integrating into market production.

In a Rwandian case study, Smit (2012: 26) describes the situation as follows:

> The departure of especially young and physically powerful man, often results in an increasing workload for staying women in the household. Besides the household duties and the care for the children, they namely have to work in agriculture and to solve all arising household problems.

---

[79]In addition to men's labor migration, (civil) wars and AIDS are also among the factors that have led to the "feminization of agriculture" in Africa. See, among others, Lastarria-Cornhiel (2008), Ajani and Igbokwe (2011).

[80]For an overview of the literature on gender-related changes in the course of deagrarianization in several African countries, see Bryceson (2019).

[81]Especially in patriarchal societies, child rearing can be made much more challenging by the absence of fathers who, in the eyes of the children, hold a position of authority and respect, which women cannot readily assume: "In patriarchal societies of Ghana, for instance, the absence of a father makes it difficult to control growing child. Thus putting emotional and psychological strain on the mothers and migrant fathers." (Migrating out of Poverty Research Programme Consortium 2016b: 2)

Some studies refer to the particular vulnerabilities of female-led households in rural areas and their economic dependence on men's remittances (e.g., Francis 2002). In a study on Niger (Niamey), Youngstedt (2013) writes, for example:

> Effectively serving as household heads to ensure social reproduction, many take on difficult burdens of extra work when their husbands fail to send remittances, while others assume responsibility for investing remittances. (Youngstedt 2013: 136)[82]

The wives left behind by migrant workers thus find themselves the de facto heads of the rural household, which corresponds to an expansion of their range of tasks. This includes, among other things, agricultural tasks: Women must often acquire—as quickly as possible—the knowledge and skills required in areas that were previously male domains.[83] Thus, a shift in the traditional distribution of gender roles is taking place *in practice* every day.

Shifting of the traditional gender distribution of labor under unchanged hierarchical relations

However, this shift in the spheres of responsibility and work does not necessarily lead to a simultaneous transformation of the hierarchical power relations within the household or community: Given that household hierarchies may exclude women from decision-making regarding migration, their taking over of 'male' work and duties may well be completely involuntary. In such cases, this extra burden of work can hardly be viewed as emancipatory (De Haas and van Rooij 2010: 59–60).[84] Furthermore, while the traditional gender-specific categorization of individual areas of work are being revised as women plow, sow, or harvest on their own, neither the women themselves nor their work efforts are receiving additional social recognition (e.g., in the form of greater decision-making authority). Thus, the gendered social order seems to remain stable in the face of these changes.[85]

---

[82]Youngstedt (2013: 136) also demonstrates that—besides the additional workload of cultivation, etc.—the women have the added duty of maintaining translocal network relationships to ensure the cohesion of the "Hausa diaspora", thereby securing an essential basis for their livelihood security.

[83]This includes, for example, the cultivation of previously "status-securing 'male' plants" [trans.] (Schäfer 2002: 32), such as millet and yams, over which previously only men had control.

[84]See Schäfer (2002).

[85]See Schäfer (2002).

Yet, Pribilsky (2004) notes that the labor migration of men can lead to a redefinition of roles in migrant families, which can positively impact partnerships:

> […] couples redefined roles and arrangements for the family life and the relationship of the partners improved due to the increased need to cooperate and coordinate despite the hardships throughout the process. (Pribilsky 2004; according to ACP 2012: 12)

Greiner and Sakdapolrak (2012) see in a case from Machakos (Kenya) another example of female empowerment through the "*feminization of agriculture*". They explain that an extreme exodus of men during the drought of the early 1980s resulted in about 60% of households being headed by women. In this context, many successful women's self-help groups were founded which, with the support of state-led soil and water campaigns, have strengthened women's rights and improved women's qualifications in these areas.[86]

Restabilization of the old gender roles through the spread of mobile communication technologies

However, in a study on rural–urban migration from Western Kenya, Ramisch (2015) reports that the old gender roles are being (re-)stabilized by the spread of mobile telecommunications: Through the extremely high frequency of telephone contact, men can establish a permanent "telecommunicative presence", meaning that women's increased autonomy over and responsibility for the rural household location is being lost again. In an effort to fulfill his role as head of the household, the absent man interferes in daily domestic and agricultural decisions without full knowledge of the actual local situation. In this way, a man can (re)consolidate his position of patriarchal power—albeit not necessarily to the benefit of the household:

> Gender relations are also put under further stress: migrant men remain (or believe they have remained) involved in rural affairs but appear to be using cellphone technologies to reinvent their household roles, replacing previously social or labor contributions with financial ones and by asserting claims over the on-farm decision-making of rural households previously considered female-headed. (Ramisch 2015: o. S.)[87]

---

[86]See Rocheleau (2001); Rocheleau and Steinberg (1995).

[87]Steel et al. (2017: 149) share this skeptical view of improved telecommunnication technologies' effects on translocal livelihoods, which they see as having "[…] added an extra layer to livelihood dynamics in the region by creating new connections as well as new dependencies. As with more physical forms of mobility, we argue that

In a nutshell: The "feminization of agriculture" resulting from the translocalization of livelihoods has significantly expanded the tasks and responsibilities of female household members in rural sub-Saharan Africa without necessarily improving their social position. And even if women make a substantial contribution to household income, for the most part, their power over these funds remains limited.

Control over the land remains problematic too: Although women make up the vast majority of the agricultural workforce, they own only a minimal share of the farmland in most African countries (FAO 2016).[88] In sub-Saharan Africa, only 15% of arable landowners are female (FAO 2011).[89] In many cases, African women's access to land impeded by the existing rights of inheritance and land use.

Women are also often disadvantaged when it comes to accessing agricultural extension services. Only 5% of smallscale farming women receive any agricultural advice (OXFAM 2013). Moreover, they experience further problems with access to small loans for investments (e.g., in agricultural technology or inputs) (FAO 2011).

The economic potential of women in agriculture is, therefore, constrained by numerous legal, policy-related and sociocultural obstacles.[90] These difficulties in accessing resources and limited resource control reduce women's overall chances of entering market production, and therewith their chances of becoming more economically autonomous and/or intensifying agricultural production.

The translocalization of livelihoods has not changed much about the general disadvantaging of women in agriculture: Gender remains the central category determining access to and control over resources in sub-Saharan Africa.

---

digital connections can be beneficial as well as disruptive for households' livelihoods."

[88]Data on women's land ownership can be found in the FAO's *Gender and Land Rights Database*: http://www.fao.org/gender-landrights-database/data-map/statistics/en/.

[89]This figure masks the wide variation between individual countries: In Mali, for example, only 5% of landowners are female, while in Botswana the figure is over 30%.

[90]One of the difficulties is the conception of small family farms under male management which persists in development cooperation and planning. Women are still rarely included in the target group of such projects (see Schäfer 2002).

6.3.3 *Health*

The connection between translocality and health is manifold. It ranges from the spread of diseases through migration and translocal lifestyles, to changes in attitudes towards health issues, to potential improvements in health care through increased financial resources or improved access to health facilities resulting from the translocal livelihoods that link rural and urban areas. It also includes the physical and mental health of both migrants and those who remain at home, as well as the impact of translocality on the overall level of health in a society.

6.3.3.1 The Influence of Migration and Translocality on the Spread of Diseases as Illustrated by HIV/AIDS

Correlation between migration and the spread of HIV

In its report on *Migration and Development in the Global South* from 2013, the IOM assumes that the mobility of people significantly accelerates the spread of infectious diseases (IOM 2013: 26). A very prominent example of this is the Ebola epidemic in West Africa (2014–2016), which developed in the border area between Guinea, Liberia and Sierra Leone and spread rapidly along the migration routes to the capitals and local urban centers. Far less contagious but much more widespread than the Ebola virus is the virus HIV, which has severely damaged parts of the African continent socially and economically. A concrete assessment of the effects of translocality on the spread of HIV and AIDS is difficult to make based on existing literature because the studies partly contradict each other. However, the majority of studies stress the connection between (circular) migration and the spread of HIV.

For example, Cremin et al. (2015) describe the correlation between migration and the spread of venereal diseases for the Gaza Province in Mozambique, which has strong seasonal patterns of circular migration connecting it to the South African mining areas in Gauteng. Generally, the miners return to their home region in December (i.e., for Christmas and end-of-year celebrations); as a result, not only does the birth rate rise there (in September of the following year), the infection rate for venereal diseases, especially HIV, rises too.[91]

Anarfi (1993) describes a similar temporal pattern for Ghana: One regular occasion for the return of migrants are local festivities (Anarfi refers to the "Apoo" of the Bono), during which otherwise prevailing moral standards are suspended.[92] For example, a man cannot denounce his wife for unfaithfulness if she sleeps with another man during this time (Anarfi 1993: 17). According to Anarfi (1993), following the infection of transmigrants via extramarital sexual intercourse at their arrival area, promiscuity during the festival leads to a more rapid spread of diseases in the region of origin.

Pison et al. (1993) also confirm this in their study on Senegal.[93] There, men mainly become infected with HIV at the destination of their migration and infect their wives or girlfriends during short stays in their home village. Lurie et al. (1997) describe a similar phenomenon among migrants in KwaZulu-Natal (South Africa):

> Female partners of male migrants clearly recognize that their partners are anything but monogamous while away. On the contrary, female partners of migrants are so certain that their partners take additional sexual partners while away that they

[91]"A seasonal pattern of births is observed in Gaza Province, a high-fertility setting, whereby there is a considerable rise in births each September. Increased frequency of conception during December is hypothesized to be related to labour migrants, including miners, returning from South Africa over the Christmas period. In Mozambique, this seasonal trend in institutional births is specific to the southern region where there is a large volume of labour migration to South Africa; it is not observed in northern provinces, where labour migration is infrequent. Given a seasonal trend of an increase in unprotected sex in December, it can be hypothesized that HIV transmission would also increase during this time period." (Cremin et al. 2015: 2)

[92]"Ghanaian migrants often visit home during festive occasions such as Christmas and local festivals, most of which are given over to revelry, in an atmosphere of laxity and permissiveness. A good example is the Apoo festival of a section of the Bono people: during the two weeks of the festival, there is a general licence to flout the existing law and order. It is reported, for example, that a man cannot divorce his wife on the grounds of adultery if another man sleeps with her during the festival period. Thus migration makes the relocation of the disease possible and the social activities facilitate its diffusion." (Anarfi 1993: 17)

[93]"HIV is mainly transmitted first to adult men through sexual contacts with infected women met during their seasonal migration and second to their wives or regular partners once they are back home." (Pison et al. 1993: 196)

> often laugh at the idiocy of such a question. This assumption is so central to the context of migration that it is barely discussed since it is recognized by all, females as well as males, as being a truism. (Lurie et al. 1997: 22)

According to studies by Konotey-Ahulu (1989) and Mabey et al. (1988), migration also has a direct influence on sexual behavior and HIV transmission in Ghana. These studies describe that two-thirds of HIV-infected prostitutes in the Gambia come from the neighboring countries of Senegal and Guinea (Mabey et al. 1988: 1) and that a large proportion of HIV infections in Ghana are related to international circular labor migration, especially from Côte d'Ivoire (Konotey-Ahulu 1989).

Abdool Karim et al. (1992) conclude in their study of HIV prevalence in rural KwaZulu-Natal that people who have recently changed their place of residence are three times more likely to be infected with HIV than "sedentary people". Nunn et al. (1995) recognize a similar pattern in their research in Uganda, where the HIV rate among permanent residents of the study area was significantly lower than among migrants. This also seems to apply to Hlabisa, a rural community in KwaZulu-Natal (South Africa): In a pilot study, Lurie et al. (1997) found that every detected case of HIV was either in a remigrant or the partner of a migrant. Zuma et al. (2003) show by means of a case study in Carltonville, South Africa, that both male migrants and female migrants are at risk:

> HIV prevalence was 37.1% with higher prevalence among migrant women (46.0%) than non-migrant women (34.7%) [...]. The highest HIV prevalence (50.9%) was between ages 26 and 35 years. (Zuma et al. 2003: 814)

Recent studies on Lesotho and Swaziland confirm the link between migrant work in South African mines and the HIV infection rate, although the correlation is not quite as strong as in the older studies:

> Using Demographic and Health Surveys, the analysis shows that migrant miners aged 30–44 are 15 percentage points more likely to be HIV positive and having a migrant miner as a partner increases the probability of infection for women by 8% points. The study also shows that miners are less likely to abstain and to use condoms and that female partners of miners are more likely to engage in extra-marital sex. (Corno and de Walque 2012: 1)

Migration accelerates the spread of AIDS in the country of origin but not the country of arrival.

A study by Docquier, Vasilakis and Tamfutu Munsi (2011), which draws on data (from the 1990s) from 44 African countries, reveals an interesting result regarding international migration's respective effects on areas of origin and arrival: While emigration significantly accelerates the spread of AIDS in the country of origin, this dynamic is not evident in the country of arrival; no measurable influence of immigration on the infection rate in the destination country was found.[94] This study makes the effect of translocal networks for the spread of the disease clear:

> However the long-run effect of emigration accounts for more than 5 percent of HIV prevalence rates in 18 countries, and more than 20 percent in 9 countries. In particular, HIV prevalence rates in the year 2000 would have been at least 20 percent larger without decreasing emigration in countries such as Mauritius, Lesotho, Swaziland, Botswana or Namibia. On the contrary, prevalence rates would have been at least 20 percent lower without increasing emigration in countries such as Burkina Faso, Comoros, Liberia or Equatorial Guinea. (Docquier et al. 2011: 3)

The connection between migration and HIV infection has lost significance due to the ubiquity of HIV/AIDS in Africa today.

The connection between the spread of HIV/AIDS and migration, which was repeatedly postulated in the late 1980s and 1990s, has lost much of its significance in the meantime. In the early 2000s, the IOM (2002: 2) observed that this connection was losing relevance for Southern Africa because AIDS had become widespread both geographically and demographically: HIV and AIDS are now ubiquitous in many parts of Africa. For South Africa, Coffee, Lurie and Garnett (2007: 344) describe the unfolding of this situation as follows:

> Early in the South African epidemic, migration appeared to fuel HIV spread by increasing geographic connectedness between

---

[94] "Our data cover 44 sub-Saharan African countries over the nineties. Controlling for omitted variables, spatial correlation and endogeneity problems, we estimate the effect of immigration and emigration on the dynamics of HIV prevalence rates. Our analysis reveals that emigration to high-prevalence destination countries increases infection rates at origin. On the contrary, immigration does not generate significant effects. Although other mechanisms are plausible, these results are consistent with the widespread view that migrants have unprotected relations with prostitutes who were already infected in the host country. Hence, immigration does not induce significant changes in prevalence rates at destination. However new infected migrants propagate the virus to their origin countries through circulation, visits and/or return migration." (Docquier et al. 2011: 3)

> locations with substantial differences in prevalence rates, such as migrant work sites and their rural home areas. In the mid-1990s, migrants in South Africa and Lesotho had levels of HIV infection close to seven times that of their non-migrant peers, and many early HIV cases were linked directly to migrant workers. However, as the epidemic progressed and prevalence rose in rural areas, the distinction between areas of high and low prevalence has blurred, and rural epidemics may be self-sustaining. (Coffee et al. 2007: 344)

Against this backdrop, the results of a recent study by Kenyon et al. (2014) are not surprising. After analyzing data from 141 countries, including 28 sub-Saharan countries, they did not find any statistical link between migration intensity (internal and international migration) and the spread of HIV: "We were not able to assess any association between HIV prevalence and the frequency of migration" (Kenyon et al. 2014: 12).[95]

Today, (translocal) migration no longer has any detectable direct influence on the spread of HIV/AIDS.

We can conclude then, that the translocalization of livelihoods in the 1990s greatly accelerated the spread of HIV, as many migrant workers became infected at their destinations and then acted as carriers of the virus, spreading it in their rural areas of origin. Meanwhile, however, (translocal) migration no longer has a direct influence on the spread of the disease.

6.3.3.2 The Effects of HIV/AIDS on (Translocal) Households

In the previous chapter, we found that HIV/AIDS spread along the migration routes and the translocal network corridors across sub-Saharan Africa but that the disease is now so widespread that it can no longer be said with certainty that translocal households are more frequently affected by AIDS than locally organized households.

AIDS and the destabilization of (translocal) household structures

Nevertheless, the AIDS pandemic poses existential problems for an extremely large number of translocal households in sub-Saharan Africa, particularly for the people residing in rural areas. On the one hand, the illness means the loss of workforce and, in the case of migrant workers who are ill, the loss of remittances which are often essential for survival. On the other hand, the disease may also lead to processes of

[95]See also Kenyon et al. (2014).

detranslocalization, as sick migrant workers often return to the rural household location to be cared for and ultimately die (*"returning home to die" phenomenon*). The return of the sick involves additional social, economic and emotional burdens[96] and often destabilizes the household. For South Africa, Clark et al. (2007: 7) describe the impact the return of sick migrants has on rural households and the healthcare system there:

> This study shows that in the context of ongoing circular labour migration in South Africa, return migration back to rural homes is associated with advancing illness and death, particularly from HIV/AIDS, which is adding an extra burden to families and health systems in rural areas. The returning-home-to-die phenomenon is associated with loss of household income (through cessation of remittances), increased household health expenditure (for healthcare and funerals), and further loss of household (and community) income given the high opportunity cost of caring for a severely ill person. There is also the effect of additional patient and cost burdens being placed on already strained rural health systems. (Clark et al. 2007: 7)

In AIDS-affected households, the children suffer, even before they are orphaned: In addition to the emotional toll children bear when parents or other adult household members show HIV-related symptoms, the children often have to take on new tasks, ranging from additional household chores (cooking, cleaning, carrying water, and laundry), to caring for the sick person or young children (feeding, washing, toilet assistance, accompanying doctors visits, etc.), to agricultural or nonagricultural work. Foster and Williamson (2000: 278) point out that in cases of parental AIDS infection, children's school attendance and performance decline significantly. This is largely because the child's workforce is needed in the household and because—owing to the loss of parental earnings and the high cost of medication—there are often no funds available for school fees.

Moreover, caring for AIDS orphans poses immense economic and social–psychological challenges[97] for translocal

---

[96]"For example, children may experience reduction in their quality of life when their mother goes to provide home care for an HIV/AIDS-affected relative or because of transfers of money to a sick relative's household." (Foster and Williamson 2000: 278)

[97]"In a Ugandan study, most children felt hopeless or angry when their parents became sick and scared their parents would die. Most orphans were depressed, with lower expectations about the future: fewer orphans

households—which have far-reaching consequences for the upbringing and education of the children and adolescents concerned.

> Early evidence for the impact on children's school attendance as a result of orphanhood came from Uganda; the financial strain led to households with orphans failing to raise funds to send their own children to school. Education is often disrupted when parents become sick, especially older girl children who are required to take over household and caregiving chores. In Uganda, amongst children 15-19 years whose parents had died, only 29% continued their schooling undisrupted; 25% lost school time and 45% dropped out of school; school-age children with the greatest chance of continuing their education were those who lived with a surviving parent; those fostered by grandparents had the least chance (7%). (Foster and Williamson 2000: 281)

The general health of AIDS orphans in rural areas is also considerably worse than that of non-orphans.[98]

Rural areas are most affected by the social consequences of AIDS. They bear the brunt of the hardship.

The logic of translocal livelihoods leads the rural areas of the African continent to be more strongly affected by the serious consequences of AIDS: It is the rural parts of translocal communities that bear much of the burden of this social disaster.

Hope is currently found in programs for antiretroviral HIV therapy (ART). Recent studies (e.g., from South Africa) have shown that ART programs help both to reduce infection rates and to increase the life expectancy of people infected with HIV.

Hope in the success of antiretroviral HIV therapy (ART).

Bor et al. (2013) have conducted research in a rural region in KwaZulu-Natal (approx. 100,000 inhabitants) where approximately half of all women and one-third of men between the ages of 35 and 49 are infected with the AIDS virus. In 2004, a large-scale program was launched to expand treatment with antiretroviral combination therapy. Bor et al. (2013) found that life expectancy has increased dramatically

---

expected to get a job, wanted to get married or wanted children than non-orphans. Depression was more likely in 10-14 years old than 15-19 years and such children were more likely to be living with a widowed father than with a widowed mother, suggesting that the loss of a mother is more distressing than the loss of a father." (Foster and Williamson 2000: 282)

[98]"Younger orphans in rural Zambia were more likely to have frequent illnesses than non-orphans." (Foster and Williamson 2000: 281)

as a result. The average life expectancy of a 15-year-old rose from 49.2 years in 2003 to 60.5 years in 2011.

Tanser et al. (2013) have carried out comparative research on the risk of infection in the same region. According to their study, the risk of infection decreased significantly with the expansion of the ART program:

> [...] an HIV-uninfected individual living in a community with high ART coverage (30 to 40% of all HIV-infected individuals on ART) was 38% less likely to acquire HIV than someone living in a community where ART coverage was low (<10% of all HIV-infected individuals on ART). (Tanser et al. 2013: 1)

Bor et al. (2013) point out that the costs for antiretroviral HIV therapy per patient range between US$500 and US$900. These costs should—we argue—be compared with the costs of the AIDS pandemic and its consequences. Treating the 25 million people infected with HIV in sub-Saharan Africa (UNAIDS 2014: 18)[99] would mean annual costs of about 17.5 billion. Considering of the apparently high effectiveness of ART and the immense social and economic costs of AIDS (e.g., through the loss of labor and the need to support the sick, widows and orphans), this figure does not seem high. For comparison: The 2014 FIFA World Cup in Brazil and the 2016 Olympic Games in Rio de Janeiro together cost significantly more (see Haferburg and Steinbrink 2017).[100]

Large-scale ART programs would simultaneously have the effect of "integrated rural development programs", as their impact is clearly multisectoral.

Since—as described above—translocal logic leads AIDS to pose a particularly *rural* "development problem" in Africa, large-scale ART programs would simultaneously have the effect of "integrated rural development programs", as their impact is clearly multisectoral (affecting education and agriculture, for example, in addition to health).

[99]Considering the number of unreported cases, the actual total is probably much higher.

[100]Another comparison: Between 2008 and 2014, German taxpayers paid a total of around US$268 billion to rescue banks. Although this measure has probably increased the life expectancy of some bankers, the money could have been used to finance a comprehensive ART program in sub-Saharan Africa for 15 years.

6.3.3.3 The Influence of Translocality on Mental Health

We are not aware of any studies on the psychosocial consequences of migration in Africa or on the influence of translocal livelihoods on the psychological health of migrants and nonmigrants. However, some conclusions can on this topic be drawn from general experiences within the African context. First, the change in the socio-spatial structure of the translocalized household results in (temporary) spatial distance, e.g., between wife and husband and/or parents and children, and thus a lower intensity of contact. Feelings of loneliness and even depression can result from a lack of emotional feedback, as can unhealthy compensation behaviors like drug or alcohol consumption. For Rwanda, Smit (2012) describes that in 33.8% of the migrant households surveyed one or more members expressed feelings of loneliness or loss as a result of migration (Smit 2012: 80).[101] Yaro et al. (2015: 124) report similar findings in a study of Ghana:

> Distance to destination constituted a major psychological challenge to the families. The NR migrant families lamented the most about missing their partners and fathers. Some even wished that the migrant abandoned the economic project and returned home to make the family complete. Marital problems emerge with the absence of these migrants, as over time both partners are accused of being unfaithful.

The long-term separation of family members can have a negative impact on psychological well-being and thus affect both mental and physical health as well as productivity in the long term, as one study from Ghana confirms (Migrating out of Poverty Research Programme Consortium 2016b: 3).[102]

---

[101]"Many more migrant households (33.8%) experience feelings of loneliness and loss as most important negative impact of migration." (Smit 2012: 80)

[102]Children face a particular emotional burden when separated from their parents. This can negatively affect school performance, as Smit (2012) reports for Rwanda: "Also several children of migrants indicated to have feelings of loneliness and sadness, resulting in a poor motivation to attend school. So, a 71-year old grandmother in Nyagasambu told: 'My granddaughter is often crying. She misses her mummy and wants to see her. I often have to explain that her mother is in the city in order to help us to survive. Every day, I stimulate her to do her best in school'." (Smit 2012: 81)

Moreover, a permanent spatial separation of partners can give rise to marital problems capable of destabilizing a translocal network (Yaro et al. 2015: 124).[103] Thus, keeping in touch over long distances is also essential when it comes to mental health (ACP 2012: 11). In this respect, improved access to mobile telecommunications and the internet can have a positive impact, as these media enable the members of translocal households to maintain a certain degree of social proximity, even over longer distances, and to support each other during phases of psychological stress (see Sterly 2015).[104]

Findings on the psychosocial aspects of translocal livelihoods are still very fragmentary, and further research is needed.

6.3.3.4 The Effects of Translocality on Health Spending and Health Consciousness

Remittances are often used for healthcare expenditures at rural household locations.

We can plausibly assume that the translocality of livelihoods both affects household spending on health-related services and medication and also leads to changes in attitudes and action with regard to health issues as a result of migrants' social remittances. However, it is unclear whether higher health expenditures reflect a greater "health awareness" in a household, or whether they are merely an expression of higher disposable income.

The remittances of urban household members are often used to finance health expenditures at rural household

[103]However, in its study in Ghana, the *Migrating out of Poverty Research Programme Consortium* (2016b: 2) reported that 24.1% of men and 15.6% of women said migration had actually positively influenced their relationship. This was attributed to the improvement in their provision situation and a lower frequency of disputes, which resulted from the geographical separation. Conversely, however, 27.6% of men and 15.6% of women stated that migration was a strain on their relationship as they had less opportunity to communicate directly with their partners.

[104]Communication via telephone (or internet) induces a feeling of closeness by lowering the threshold of contact and thereby promoting its frequency. This closeness is essential for maintaining social and emotional bonds, e.g., between couples living in separate places, parents and children, siblings, friends, etc. Feeling more connected to one's native home can alleviate psychological pressure and feelings of loneliness, contributing positively to mental and physical health. In Niamey (Niger), for example, 85% of Hausa migrants use mobile phones to stay in touch with their families (Youngstedt 2013: 149).

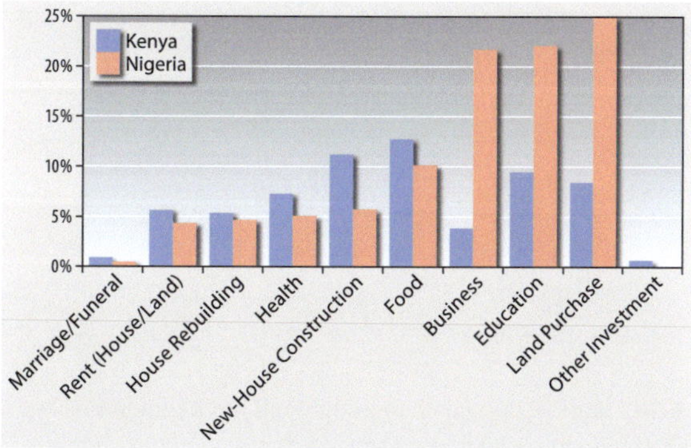

Fig. 6.9 Reported use of international remittances: Kenya and Nigeria (as a percentage of total). *Source* Modified from Watkins and Quattri (2014: 15); with data from Ratha et al. (2011: 20)

locations. Based on the available studies, however, it is difficult to quantify the share of such expenditures in the total volume of remittances, not least because of the lack of reliable aggregated data on remittances from internal migrants (see Sect. 6.1.1). For this reason, the sources only allow us to make some estimates for international remittances:

International transmigration leads to an improvement in the health care or health status of relatives in the area of origin.

With regard to Kenya and Nigeria, Watkins and Quattri (2014: 15) show that between 5% (Nigeria) and 7% (Kenya) of international remittances are spent on health care (see Fig. 6.9).

Although these percentages seem quite small compared to spending areas like land purchase, education and business, in absolute terms they are likely to make a difference. Indeed, Zourkaleini et al. (2013) observe for Cameroon that expenditures on health in households with international migrants (including remigrants) are significantly higher than in households without migrants (with the highest values among households that currently have a relative abroad [according to IOM 2013: 30]). To this extent, we can assume that international migration leads to an improvement in the health care or health status of relatives in the area of origin.

However, the question remains whether these values can be extrapolated from international transmigration to domestic translocal migration. As previously described in Sect. 6.1.1.3, households with international migrants and those with domestic migrants show significant differences in prosperity levels, which affect spending patterns—including in the health sector.

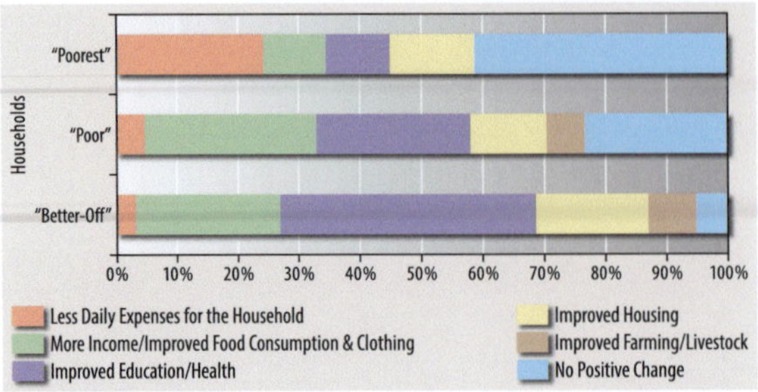

Fig. 6.10 Most important changes through the migration of household members (by socioeconomic status; n = 130). *Source* Modified from Smit (2012: 108)

Regarding domestic remittances, Smit (2012) states that over 40% of migrant households in a rural region in Rwanda reported using remittances for "health"; only the area of "nutrition" was mentioned more frequently (see Smit 2012: 100 and Fig. 6.7 in Sect. 6.1.1.3). However, in absolute values, the amount spent on health is relatively small.[105] Nevertheless, 43% of the migrant households that Smit surveyed around Kigali reported experiencing an improvement in the area of health as a result of migration; a large proportion specifically stated that the remittances enabled them to afford health insurance, which would provide them with the necessary medications and medical services in the event of illness (Smit 2012: 104). Here, however, the socioeconomic status of the migrant household is decisive:

> Better-off migrant households namely experience significantly more often a positive change in access to health care than poorer migrant households. (Smit 2012: 105)

Particularly enlightening in this respect are how the 130 rural migrant households that Smit (2012) surveyed responded to the question in which areas they have experienced the most important changes as a result of migration. Here, considerable discrepancies between different wealth groups are evident (see Fig. 6.10).

---

[105]Most of the money that flows into rural areas in the form of remittances is used to improve the housing situation and nutrition of rural households (see Smit 2012: 100 as well as Figs. 6.6 and 6.7 in Sect. 6.1.1.3)

Socioeconomic stratification is also manifested in the health outcomes—both in local and translocal contexts.

The most common response from the poorest households: Migration had not led to any significant changes. And those households that saw improvements mostly mentioned that their daily expenses declined. With increasing prosperity—the results show—this aspect recedes into the background. Wealthier households most often mention improvements in the areas of "education" and "health".

Overall, it becomes clear that socioeconomic stratification is also manifested in the health outcomes—both in local and translocal contexts. For some of the translocal households, remittances can be expected to result in very concrete improvements in medical treatment and preventive care (e.g., in the form of health insurance); however, for many vulnerable households, remittances—if they receive any—only serve to meet the most basic needs.

In their study on *social remittances* in Tanzania, Helgesson Sekei et al. (2014: 59) show that migration experiences can change attitudes and practices regarding health. Any change in health awareness is most likely to occur among the migrants themselves, but because of their embeddedness in translocal communication networks, this "new" knowledge also reaches their relatives and friends living in rural areas of origin. It is reasonable to assume that with the widespread use of mobile telephones today, the exchange of knowledge and experience regarding health-relevant issues and health-related decision-making processes in translocal households are different now than they were before (Ramisch 2015: see above). To our knowledge, however, the influence of mobile telephony on the health situation of translocal households has not yet been investigated.

Nor do we know whether translocal networks are already being used for public health and medical prevention work. It seems that it would be advantageous to use the existing complex and geographically far-reaching translocal channels of informal communication, for example, to promote large-scale *health awareness campaigns* and to involve circular migrants as influencers in health education.

6.3.3.5 The Influence of Translocality on Access to Nursing and Medical Care

The impact of translocal livelihood organization on household members' access to nursing and medical care in old age or cases of illness is not only related to the economic prerequisites of care. The social and spatial structure of translocal networks also constitutes an important condition for such care.

Translocal networks enable the elderly and the sick to move to make use of the medical care and/or nursing facilities available at different locations.

As discussed earlier, the movement of people within established translocal networks between rural and urban areas comprises more than just labor migration; old and sick people are also mobile within the network corridors.[106] The reason for this is that each of the places linked by translocal networks has different social and infrastructural opportunity structures concerning health care. Translocal networks enable the elderly and the sick to move to make use of the medical and/or nursing care available at the different localities, as necessary. The translocal mobility of elderly and sick people, therefore, takes place in both directions between rural and urban areas for different reasons:

The female members at rural household locations are usually responsible for the "care role".

Several times in this report we have pointed out that many migrants move back from the city to the countryside at the end of their working lives or when they are no longer able to work (see "*retirement migration*" [Sect. 5.3.4.2] and the "*returning home to die phenomenon*" [Sect. 6.3.3.2]). This migration can be explained in economic terms (reduction of the cost of living, which is generally lower in rural areas) or against the backdrop of cultural norms (e.g., ancestral beliefs).

Age- and disease-related mobility from the region of origin to urban centers is mostly due to the uneven spatial distribution of medical care facilities.

In this context, it is important to note that—in accordance with the translocal system of labor division and cooperation—the household members in rural areas, primarily women, are responsible for providing care ("*care role*") (see Greiner 2008, 2010, Lohnert 2002; Schmidt-Kallert 2009: 326–327). Age- or disease-related remigration thus offers (labor) migrants the opportunity for social support, which is often unavailable at urban household locations. Health migration from the city to the countryside can also be aimed at accessing traditional healing methods in the village of origin. In some cases, this involves the availability of specific remedies or the consultation of certain healers (see Steinbrink 2009, among others); another reason that sick people may return to their region of origin is to "consult" the ancestors.

Age- and disease-related mobility from the region of origin to urban centers is mostly attributable to the uneven

[106]See also the *Phase Model of Translocalization*, Phase 3, Sect. 5.3.3.

spatial distribution of medical care facilities.[107] In many rural areas of sub-Saharan Africa, the provision of medical care is hardly ensured; there are too few doctors, hospitals, and health centers, and the existing medical infrastructure is usually extremely poorly equipped. Even though conditions in most African cities are also highly problematic, they are generally better than in rural areas. This is why people use translocal (intra-household) networks to gain (better) access to the medical infrastructure and care available in cities. This encompasses acute cases of illness as well as long-term and preventive treatments, including the pre- and postnatal care of mothers for example. The urban household location offers the health migrant accommodation, and the relatives living there provide the necessary social and financial support during the treatment period. Steinbrink (2009) reports from South Africa that elderly people also move from the countryside to the city to receive certain medical examinations and services to which they have little or no access in the countryside. People also temporarily move out of rural areas to apply for *disability grants*, because the related medical examinations and bureaucratic procedures are easier to organize in urban centers, like Cape Town, than in peripheral rural areas.

In some cases, however, health migration is linked to other migration motives. For example, Peil, Ekpenyong and Oyeneye (1988) report that older widows from southern Nigeria often move to cities to look after the children of their sons working there; this (a) ensures the widows' retirement and (b) grants the older women proximity to better medical facilities and the "easier life" in the city.

Translocal networks are essential to formal and informal care systems in many African societies.

Overall, we can infer that translocal networks are essential for the functioning of formal and informal care systems in many African societies. In the event of illness, much the urban population is dependent on the informally organized nursing care provided by relatives in rural areas; and it is also through translocalized social networks that people in rural areas are able to access healthcare infrastructure in the city. To the best of our knowledge, however, the importance of translocality for health care in Africa has not yet been examined in detail. Here, too, there is a need for research.

[107]Even in relatively prosperous South Africa, the distribution of medical care facilities remains extremely unequal (see e.g., Tanser et al. 2006 and Ijumba et al. 2004).

References

Abdool Karim Q, Abdool Karim SS, Singh B, Short R, Ngxongo S (1992) Seroprevalence of HIV infection in rural South Africa. AIDS 6(12):1535–1539

Abdulkadir H, Abdullahi A (2018) Factors influencing women access to land in Bichi local. Int J Inform Res Rev 5(4):5395–5398

Abdul-Korah GB (2011) 'Now if you have only sons you are dead': Migration, gender, and family economy in twentieth century northwestern Ghana. J Asian Afr Stud 46(4):390–403

ACP (2012) Transnational families and the social and gender impact of mobility in ACP countries. Background Note: ACPOBS/BN06

Adepoju A (2005) Migration in West Africa. Global Commission on International Migration. Human Resources Development Centre, Lagos

Adepoju A (2008) Migration in Subsaharan Africa. The Current African Issues 37. Nordiska Afrikainstitutet, Uppsala

Adger NW (2006) Vulnerability. Glob Environ Change 16:264–281

Afifi T (2009) Niger: case study report. EACH-FOR environmental change and forced migration scenarios. Available via EACH-FOR. http://rosamartinez.org/wp-content/uploads/2015/11/Migraciones-y-Cambio-Climatico_EACHFOR.pdf. Accessed 18 Mar 2019

Ajani EN, Igbokwe EM (2011) Implications of feminization of agriculture on women farmers in Anambra State Nigeria. J Agric Ext 15(1):31–39

Ajanovic E (2014) Remittances—More than money? Konzeptionelle Überlegungen zum Verständnis von Rücküberweisungen von MigrantInnen in ihre Sendeländer. In: Ataç I, Fanizadeh M, Kraler A, Manzenreiter W (eds) Migration und Entwicklung. Neue Perspektiven, Wien, pp 101–116

Aklilu Y, Catley A (2010) MIND THE GAP. Commercialization, Livelihoods and Wealth Disparity in Pastoralist Areas of Ethiopia. Feinstein International Centre

Ambrosius C, Fritz B, Stiegler U (2008) Geldsendungen von Migranten —‚Manna' für die wirtschaftliche Entwicklung? GIGA German Institute of Global and Area Studies—Leibniz-Institut für Globale und Regionale Studien, Hamburg

Anarfi JK (1993) Sexuality, migration and AIDS in Ghana—a socio-behavioural study. Health Trans Rev 3:45–67

Anker R, Hein C (1986) Introduction and overview. In: Anker R, Hein C (eds) Sex inequalities in urban employment in the third world. St. Martin`s Press, New York, pp 1–56

Aufenvenne P, Felgentreff C (2013) Umweltmigranten und Klimaflüchtlinge – zweifelhafte Kategorien in der aktuellen Debatte. IMIS-Beiträge 44:19–44

Aufenvenne P, Felgentreff C, Heller W (2013) Klimaflucht und Umweltmigration – diskussionswürdige Begriffe. In: Reilich J, Frohwerk S (eds) Eine Sonate der Ökonomie – Sätze zur allgemeinen Theorie, der Raumwirtschaft und der Klimapolitik. Festschrift für Klaus Schöler zum 65. Geburtstag. S. Roderer-Verlag, Magdeburg, pp 25–51

Awumbila M (2015) Women moving within borders. Gender and internal migration dynamics in Ghana. Ghana J Geogr 7(2):132–145

Awumbila M, Darkwah A, Teye J (2015) Migration, intra-household dynamics and youth aspirations. Unpublished Paper. Migrating out of Poverty, University of Sussex

Ayana Aga G, Martinez P, Soledad M (2014) International remittances and financial inclusion in Sub-Saharan Africa. In: Policy Research Working Paper 6991, World Bank

Baker J (1994) Small urban centres and their role in rural restructuring. In: Zegeye A, Pausewang S (eds) Ethiopia in change. Peasantry, Nationalism and Democracy. British Academic Press, London/New York, pp 152–171

Barrett C, Reardon T, Webb P (2001) Nonfarm income diversification and household livelihood strategies in Rural Africa: concepts dynamics and policy implications. Food Policy 26(4):315–331

Batista C, Lacuesta A, Vicente PC (2010) Testing the ‘Brain Gain’ hypothesis: micro evidence from cape verde. J Dev Econ 97(1):32–45

Beguy D, Bocquier P, Zulu EM (2010) Circular migration patterns and determinants in Nairobi slum settlements. Demogr Res 23:549–586

Black R, Bennett S, Thomas SM, Beddington JR (2011) Climate change: migration as adaptation. Nature 478(7370):447–449

Blasche M, Inhetveen H (1983) Frauen in der kleinbäuerlichen Landwirtschaft. Westdeutscher Verlag, Opladen

Bodin Ö, Crona B, Ernstson H (2005) Social networks in natural resource management: what is there to learn from a structural perspective? Available via Ecology and Society. http://www.ecologyandsociety.org/vol10/iss1/resp2/. Accessed 18 Mar 2019

Bogardi J, Warner K (2009) Here comes the flood. Nat Reports Clim Change 3:9–11

Bor J, Herbst AJ, Newell ML, Bärnighausen T (2013) Increases in adult life expectancy in rural South Africa; valuing the scale-up of HIV treatment. Science 339(6122):961–965

Brockerhoff M, Eu H (1993) Demographic and socioeconomic determinants of female rural to urban migration in Sub-Saharan Africa. Int Migrat Rev 27(3):557–577

Bryceson DF (2019) Gender and generational patterns of African deagrarianization: evolving labor and land allocation in smallholder peasant household farming, 1980–2015. World Dev 113:60–72

Calì M, Cantore N (2010) The impact of circular migration on source countries: a simulation exercise. Overseas Development Institute

Carr ER (2005) Placing the environment in migration: environment economy and power in Ghana’s central region. Environ Plan A: Econ Space 37(5):925–946

Castaldo A, Deshingkar P, McKay A (2012) Internal migration, remittances and poverty: evidence from Ghana and India. University of Sussex. Available via Migrating out of Poverty. http://migratingoutofpoverty.dfid.gov.uk/files/file.php?name=wp7-internal-migration-remittances-and-poverty.pdf&site=354. Accessed 18 Mar 2019

Clark SJ, Collinson MA, Kahn K, Drullinger K, Tollman SM (2007) Returning home to die: circular labour migration and mortality in South Africa. Scand J Public Health 69:35–44

Coffee M, Lurie MN, Garnett GP (2007) Modelling the impact of migration on the HIV epidemic in South Africa. AIDS 21(3):343–350

Conelly T (1994) Population pressure, labor availability, and agricultural disintensification: the decline of farming on Rusinga Island, Kenya. Human Ecol 22(2):145–170

Corno L, de Walque D (2012) Mines migration and HIV/AIDS in Southern Africa. J Afr Econ 21(3):465–498

Cottyn I, Schapendonk J (2013) Mobility in Sub-Saharan Africa: patterns, processes and policies. Rurban Africa: State of the Art Report 2, Department of Geosciences and Natural Resource Management, University of Copenhagen

Cotula L, Toulmin C (eds) (2004) Till to tiller: international migration, remittances and land rights in West Africa. International Institute for Environment and Development, London

De Haan A (2000) Migrants, livelihoods, and rights: the relevance of migration in development policies. Social Development Working Paper 4, London

De Haan A (2006) Migration in the development studies literature: has it come out of its marginality? Research Paper 19, United Nations University

De Haas H (2010) Migration and development: a theoretical perspective. Int Migrat Rev 1(44):227–264

De Haas H, van Rooij A (2010) Migration as emancipation? The impact of internal and international migration on the position of women in rural Morocco. Oxf Dev Stud 38:43–62

Deshingkar P (2004) Understanding the implications of migration for pro-poor agricultural growth. Overseas Development Institute

Dilger H (2013) Securing wealth, ordering social relations: kinship, morality, and the configuration of subjectivity and belonging across the rural-urban divide. In: Kane A, Leedy T (eds) African migrations: patterns and perspectives. University Press, Bloomington, pp 113–132

Dinan C (1983) Sugar daddies and gold-diggers: the white-collar single women in Accra. In: Oppong C (ed) Female and male in West Africa. George Allen and Unwin, London, pp 344–366

Docquier F, Vasilakis Ch, Tamfutu Munsi D (2011) International migration and the propagation of HIV in Sub-Saharan Africa. Afr J Health Econ 35:20–33

Dorlöchter-Sulser S (2014) Wandel von Livelihood-Systemen im Spannungsfeld von Struktur und Handeln, eine historisch angelegte Livelihood-Analyse von 1960 bis 2010 am Beispiel der Region Dosso, Niger. Wvb, Berlin

Doss C, McPeak J, Barrett C (2006) Interpersonal, intertemporal and spatial variation in risk perceptions: evidence from East Africa. World Dev 36(8):1453–1468

Ekbom A, Knutsson P, Ovuka M (2001) Is sustainable development based on agriculture attainable in Kenya? A multidisciplinary case study of Murang'a district land. Degrad Dev 12(5):435–447

Etzold B (2017) Mobility, space and livelihood trajectories: new perspectives on migration, translocality and place-making for livelihood studies. In: De Haan L (ed) Livelihoods and development: new perspectives. Brill, Leiden, pp 44–68

Evans HE, Ngau P (1991) Rural-urban relations, household income diversification and agricultural productivity. Dev Change 22:519–545

FAO (2003) Gender—Key to sustainability and food security. Available via FAO. http://www.fao.org/3/w4430e/w4430e00.pdf. Accessed 18 Mar 2019

FAO (2011) The state of food and agriculture, women in agriculture, closing the gender gap for development. Available via FAO. http://www.fao.org/3/a-i2050e.pdf. Accessed 18 Mar 2019

FAO (2016) Gender and land statistics. Available via FAO. http://www.fao.org/3/a-i5488e.pdf. Accessed 18 Mar 2019

Findley SE (1994) Does drought increase migration? A study of migration from rural Mali during the 1983–1984 drought. Int Migrat Rev 28(3):539–553

Folke C (2006) Resilience: the emergence of a perspective of social-ecological systems analyses. Glob Environ Change 16:253–267

Folke C, Carpenter S, Walker B, Scheffer M, Chapin T, Rockstrom J (2010) Resilience thinking: integrating resilience adaptability and transformability. Ecol Soc 15(4):20

Foster G, Williamson J (2000) A review of current literature of the impact of HIV/AIDS on children in Sub-Saharan Africa. AIDS 14 (3):275–284

Francis E (2002) Gender migration and multiple livelihoods: cases from eastern and southern Africa. J Dev Stud 38(5):167–190

Gallopín G (2006) Linkages between vulnerability resilience and adaptive capacity. Glob Environ Change 16:293–303

Godoy J, Tortora B, Sonnenschein J, Kendall J (2012) Payments and money transfer behavior of Sub-Saharan Africans. Bill and Melinda Gates Foundation, GALLUP. Available via findevgateway. https://www.findevgateway.org/sites/default/files/mfg-en-paper-payments-and-money-transfer-behavior-of-sub-saharan-africans-jun-2012.pdf. Accessed 18 Mar 2019

Greiner C (2008) Zwischen Ziegenkraal und Township: Migrationsprozesse in Nordwest-Namibia. Dietrich Reimer, Berlin

Greiner C (2011) Migration, translocal networks and socio-economic stratification in Namibia. Africa 81(4):606–627

Greiner C, Peth S, Sakdapolrak P (2014) Deciphering migration in the age of climate change: towards an understanding of translocal relations in social-ecological systems. In: TransRe Working Paper, vol 2, Department of Geography, University of Bonn, Bonn

Greiner C, Sakdapolrak P (2012) Rural-urban migration, agrarian change, and the environment in Kenya: a critical review of the literature. Popul Environ 34(4):524–553

Grosz-Ngaté M (1991) Gender generation and power labor migration as a terrain of contest in rural Mali. Institute for African Development, Cornell University

Gugler J, Ludwar-Ene G (1995) Gender and migration in Africa South of the Sahara. In: Baker J, Aina TA (eds) (1995) The migration experience in Africa. Nordiska Afrikainstitutet, Sweden, pp 257–269

Guigou B, Leriollais A (1992) Crise de l'agriculture et marginalization économique des femmes sereer siin (Sénégal). Sociétés Espaces Temps 1:45–64

Haferburg C, Steinbrink M (2017) Mega-events in emerging nations and the festivalization of the urban backstage. The cases of Rio de Janeiro and South Africa. In: Hannigan J, Richards G

(eds) Handbook of new urban studies. Los Angeles, SAGE, pp 267–290
Hahn HP (2004) Zirkuläre Arbeitsmigration in Westafrika und die, Kultur der Migration‘. Afr Spectr 39:381–404
Hamer A (1981) Diola women and migration: a case study. In: Colvin LG, Ba C, Barry B, Faye J (eds) The uprooted of the Western Sahel. Praeger, New York, pp 163–203
Helgesson Sekei L, Alvater A, Mrema J C, Kisinda A (2014) Back home: exploring the potential of South-South social remittances in the United Republic of Tanzania. Research Report, Brussels, ACP
Hollos M (1991) Migration, education and the status of women in Southern Nigeria. Am Anthropol 93:852–870
Hummel D (2015) Gendereinflüsse und soziale Differenzierungen der klimabedingten Migration. Available via Klimanavigator. http://www.klimanavigator.de/dossier/artikel/057756/index.php. Accessed 18 Mar 2019
IHSN Survey Catalog (2005) Ghana living standard survey. Available via IHSN. http://catalog.ihsn.org/index.php/catalog/63/study-description. Accessed 18 Mar 2019
Ijumba P, Day C, Ntuli A (2004) A South African health review 2003/2004. Health System Trust, Durban
IOM (2002) Labour migration and HIV/AIDS in Southern Africa. International Organization for Migration, Regional Office for Southern Africa. Available via Researchgate. https://www.researchgate.net/publication/241344885_Labour_migration_and_HIVAIDS_in_Southern_Africa. Accessed 18 Mar 2019
IOM (2010) The future of migration: building capacities for change. International Organization for Migration-World Migration Report 2010. Available via IOM. https://www.iom.int/world-migration-report-2010. Accessed 18 Mar 2019
IOM (2013) Migration and development within the south: new evidence from African Caribbean and Pacific countries. International Organization for Migration-Migration Research Series 46. Available via IOM. https://publications.iom.int/books/migration-and-development-within-south-new-evidence-african-caribbean-and-pacific-countries. Accessed 18 Mar 2019
Jónsson G (2010) The environmental factor in migration dynamics—a review of African case studies. International Migration Institute Working Paper 21, University of Oxford
Kenyon C, Buyce J, Colebunders R (2014) Classification of incidence and prevalence of certain sexually transmitted infections by world regions. Int J Infect Dis 18:73–80
KNOMAD (2016) Migration and remittances: recent developments and outlook. Migration and Development Brief 26
Konotey-Ahulu F (1989) What is AIDS? Tetteh-A'Domeno Company, Worcester
Langthaler M, Scharer V (2005) Bildungsökonomisierung in den Entwicklungsländern: Formen, Auswirkungen und Implikationen für die Bildungszusammenarbeit, Working Paper, No. 6, Austrian Foundation for Development Research (ÖFSE), Vienna
Lastarria-Cornhiel S (2008) Feminization of agriculture: trends and driving forces. Working Paper 08, World Development Report Background Papers. Available via World Bank. http://siteresources.worldbank.org/INTWDRS/Resources/477365-1327599046334/

8394679-1327599874257/LastarriaCornhiel_FeminizationOfAgri.pdf. Accessed 18 Mar 2019

Levitt P (1998) Social remittances: migration driven local-level forms of cultural diffusion. Int Migr Rev 32(4):926–948

Levitt P, Lamba-Nieves D (2011) Social remittances revisited. J Ethn Migr Stud 37(1):1–22

Lewis LA (1985) Assesing soil loss in Kiambu and Murang'a Districts Kenya. Geogr Ann 67(3/4):273–284

Litchfield J, Rolla P, Jena F, Dzingirai U, Nyikahadzoi K, Mutopo P (2018) Migrant remittances and gender in Zimbabwe. Migrating Out of Poverty Working Paper. Migrating Out of Poverty Research Programme Consortium, University of Sussex

Lohnert B (2002) Vom Hüttendorf zur Eigenheimsiedlung: Selbsthilfe im städtischen Wohnungsbau: Ist Kapstadt das Modell für das neue Südafrika?. Universitätsverlag Rasch, Osnabrück

Lopez-Ekra S, Aghazarm C, Kötter H, Mollard B (2011) The impact of remittances on gender roles and opportunities for children in recipient families: research from the international organization for migration. Gender Dev 19(1)

Ludwar-Ene G (1986) Explanatory and remedial modalities for personal misfortune in a West African society. Antropos 81:555–565

Lurie M, Harrison A, Wilkinson D, Abdool Karim S (1997) Circular migration and sexual networking in rural KwaZulu/Natal: implications for the spread of HIV and other sexually transmitted diseases. Health Trans Rev 7(3):17–27

Mabey DCW, Tedder RS, Hughes ASB, Corrah PT, Goodison SJF, O′ Connor T, Shenton FC, Lucas SB, Whittle HC, Greenwood BM (1988) Human retroviral infections in the Gambia prevalence and clinical features. Br Med J 296(6615):83–86

McKay A, Deshingkar P (2014) Internal remittances and poverty: further evidence from Africa and Asia. Working Paper 12: Migrating out of Poverty, University of Sussex

McLeman R, Smit B (2006) Migration as an adaptation to climate change. Clim Change 76:31–53

Migrating out of Poverty Research Programme Consortium (2016a): Migration into cities in Ghana: the economic benefits to migrants and their households. Available via migratingoutofpoverty. http://migratingoutofpoverty.dfid.gov.uk/files/file.php?name=2015-16-3-cms-rp13-counterf-migration-into-cities-economic-benefits-final-bfp.pdf&site=354. Accessed 18 Mar 2019

Migrating out of Poverty Research Programme Consortium (2016b): Social benefits and losses of migrating into cities in Ghana. Available via migratingoutofpoverty. http://migratingoutofpoverty.dfid.gov.uk/files/file.php?name=2015-16-4-cms-rp13-counterf-migration-into-cities-social-benefits-and-losses-final-bfp.pdf&site=354. Accessed 18 Mar 2019

Mims C (2013) 31% of Kenya's GDP is spent through mobile phones. Available via Quartz. https://qz.com/57504/31-of-kenyas-gdp-is-spent-through-mobile-phones/. Accessed 18 Mar 2019

Misiko M (2007) Fertile ground? Soil fertility management and the African smallholder. PhD-thesis, Wageningen University

Morrissey J (2011) Rethinking the 'debate on environmental refugees' from 'maximilists and minimalists' to 'proponents and critics'. J Polit Ecol 19:36–49

Mortimore M, Tiffen M (2004) Introducing research into policy: lessons from district studies of dryland development in Sub-Saharan Africa. Dev Policy Rev 22(5):259–285

Mothobi O, Grzybowski L (2017) Infrastructure deficiencies and adoption of mobile money in Sub-Saharan Africa. Inf Econ Policy (40):71–79

MPI (Migration Policy Institute) (2012) The global remittances guide. Available via MPI Data Hub. https://www.migrationpolicy.org/programs/data-hub/global-remittances-guide. Accessed 18 Mar 2019

Myers N (2002) Environmental refugees: a growing phenomenon of the 21st century. Philos Trans B R Soc 357(1429):609–613

Niedenführ, H (2018) Kindermigration in Burkina Faso: Die Praxis der "Anvertrauung" (confiage). Available via Bundeszentrale für politische Bildung. http://www.bpb.de/gesellschaft/migration/kurzdossiers/268029/kindermigration-in-burkina-faso. Accessed 17 Apr 2019

Nunn AJ, Wagner H-U, Kamali A, Kengeya-Kayondo JF, Mulder DW (1995) Migration and HIV-1 seroprevalence in a rural Ugandan population. AIDS 9(5):503–506

Nyangena W (2006) Essays on soil conservation, social capital and technology adoption. Available via Göteborg University. https://gupea.ub.gu.se/bitstream/2077/2916/1/Nyangena_full.pdf. Accessed 18 Mar 2019

O'Keefe P (1983) The causes, consequences and remedies of soil erosion in Kenya. Ambio 12(6):302–305

Obokata R, Veronis L, McLeman R (2014) Empirical research on international environmental migration: a systematic review. Popul Environ 36:111–135

OECD (Organization for Economic Co-Operation and Development) (2009) International migration outlook SOPEMI Report 2009. Available via OECD. https://www.oecd.org/migration/48342373.pdf. Accessed 18 Mar 2019

OECD (Organization for Economic Co-Operation and Development) (2010) development aid reaches an historic high in 2010. Available via OECD. http://www.oecd.org/dac/stats/developmentaidreachesanhistorichighin2010.html. Accessed 18 Mar 2019

Offiong DA (1982) The 1978–79 AkpanEkwong anti-witchcraft crusade in Nigeria. Anthropologica 24(1):27–42

Oliver-Smith A (2012) Debating environmental migration: society, nature and population displacement in climate change. J Int Dev 24:1058–1070

Olson JM, Atieno F, Muchugu E (2004) Multi-scale analysis of land use and management change on the eastern slopes of Mt. Kenya. LUCID Project Working Paper 20, Michigan State University

Oltmer J (2015) Zusammenhänge zwischen Migration und Entwicklung. Available via Deutsche Welthungerhilfe e. V. https://www.welthungerhilfe.de/aktuelles/publikation/detail/zusammenhaenge-zwischen-migration-und-entwicklung/. Accessed 18 Mar 2019

Ouédraogo J-B (1995) The girls of Nyovuruu Dagara female labour migrations to Bobo-Dioulasso. In: Baker J, Akin Aida T (eds) The migration experience in Africa. Nordiska Afrikainstitutet, Uppsala, pp 303–320

OXFAM (2013) Promises, power and poverty—corporate land deals and rural women in Africa. Available via OXFAM. http://www.

wluml.org/sites/wluml.org/files/bp170-promises-power-poverty-land-women-090413-en.pdf. Accessed 18 Mar 2019
Peil M, Ekpenyong SK, Oyeneye OY (1988) Going home: migration careers of Southern Nigerians. Int Migrat Rev 22:563–585
Peil M, Sada PO (1984) African urban society. Wiley, Chichester, New York
Pelling M, High C (2005) Understanding adaptation: what can social capital offer assessments of adaptive capacity? Glob Environ Change 15(4):308–319
Pickbourn L (2011) Migration remittances and intra-household allocation in Northern Ghana: does gender matter? Doctoral Dissertations, University of Massachusetts Amherst. Available via Scholarworks. https://scholarworks.umass.edu/dissertations/AAI3482654/. Accessed 18 Mar 2019
Piguet E (2013) From 'primitive migration' to 'climate refugees': the curious fate of the natural environment in migration studies. Ann Assoc Am Geogr 103(1):148–162
Pison G, Guenno BL, Lagarde E, Enel C, Seck C (1993) Seasonal migration: a risk factor for HIV infection in rural senegal. J Acquir Immune Defic Syndr 6(2):196–200
Pittin R (1984) Migration of women in Nigeria: the hausa case. Int Migrat Rev 18(4):1293–1314
Pott A, Felgentreff C (2016) Climatic turn in migration studies? Geographical perspectives on the relationship between climate and migration. Erde 147(2):73–80
Pribilsky J (2004) 'Aprendemos a Convivir': conjugal relations, co-parenting and family life among ecuadorian transnational migrants in New York City and the Ecuadorian Andes. Glob Netw 4(3):313–334
Quisumbing A (2003) Household decisions, gender and development. A synthesis of recent research. John Hopkins University Press for International Food Policy Research Institute (IFPRI), Baltimore
Rain D (1999) Eaters of the dry season: circular labor migration in the West African Sahel. Westview Press, Boulder
Ramisch J (2014) 'We will not farm like our fathers did' multilocational livelihoods cellphones and the continuing challenge of rural development in Western Kenya. In: Sick D (ed) Rural livelihoods regional economies and processes of change. Routledge, London, pp 10–25
Ramisch J (2015) 'Never at Ease': cellphones multilocational livelihoods and the metabolic rift in Western Kenya. Agric Hum Values 33(4):979–995
Ratha D, Mohapatra S, Özden C, Plaza S, Shaw W, Shimeles A (2011) Leveraging migration for Africa: remittances skills and investments. Available via World Bank Group. https://siteresources.worldbank.org/EXTDECPROSPECTS/Resources/476882-1157133580628/AfricaStudyEntireBook.pdf. Accessed 18 Mar 2019
Ratha D, Riedberg J (2005) On Reducing Remittance Costs. Available via World Bank. http://www.rrojasdatabank.info/iadbremit/reducing05.pdf. Accessed 18 Mar 2019
Reardon T (1997) Using evidence of household income diversification to inform study of the non-farm labour market in Africa. World Dev 25(5):735–748

Rocheleau DE (2001) Complex communities and relational webs uncertainty, surprise and transformation in Machakos. IDS Bull 32 (4):78–87

Rocheleau DE, Steinberg PE (1995) Environment development crisis and crusade: Ukambani Kenya, 1890–1990. World Dev 23(6):1037–1051

Rodima-Taylor D, Olwig MF, Chhetri N (2012) Adaptation as innovation, innovation as adaptation: an institutional approach to climate change. Appl Geogr 33:107–111

Sakdapolrak P, Naruchaikusol S, Ober K, Peth S, Porst L, Rockenbach T, Tolo V (2016) Migration in a changing climate. Towards a translocal social resilience approach. Die Erde 147(2):81–94

Sander C, Munzele Maimbo S (2005) Migrant labour remittances in Africa: reducing obstacles to developmental contributions Africa region. Working Paper Series 64. Available via The World Bank. http://documents.worldbank.org/curated/en/191171468742795861/pdf/276770PAPER0AFR0wp64.pdf. Accessed 18 Mar 2019

Schade J (2013) Entitlements capabilities and human rights. In: Faist T, Schade J (eds) Disentangling migration and climate change: toward an analysis of methodologies. Political discourses and human rights. Springer, Dordrecht/Heidelberg/New York/London, pp 201, 231–253

Schäfer R (2002) Gender und ländliche Entwicklung in Afrika. Aus Politik und Zeitgeschichte B 13–14:31–38

Scharwatt C, Williamson C (2015) Mobile money crosses borders: new remittance models in West Africa. Available via GSMA. https://www.gsma.com/mobilefordevelopment/wp-content/uploads/2015/04/2015_MMU_Mobile-money-crosses-borders_New-remittance-models-in-West-Africa.pdf. Accessed 18 Mar 2019

Scheffran J, Marmer E, Sow P (2011) Migration as a contribution to resilience and innovation in climate adaptation: social networks and co-development in Northwest Africa. Appl Geogr 1–9

Schmidt-Kallert E, Kreibich V (n d) Informelle Stadt-Land-Beziehungen. Universität Dortmund

Schutten M (2012) Rwanda livelihood diversity: causes of rural-urban migration; Why Rwanda poverty classification does not explain migration flows. Master Dissertation, University of Utrecht and National University of Rwanda

Seibert J (1995) Lebensstrategien unabhängiger junger Frauen in Lomé, Togo. Dissertation Thesis, Universität Bayreuth

Shisanya CA, Khayesi M (2007) How is climate change perceived in relation to other socioeconomic and environmental threats in Nairobi, Kenya. Climate Change 85(3):271–284

Singlemann J (1993) Levels and trends of female internal migration in developing countries, 1960–80. UN, Internal Migration of Women in Developing Countries, pp 77–93

Smit A (2012) Impacts of rural-urban migration on rural migrant households in the surroundings of Kigali. Master Thesis, University of Utrecht

SOFA Team and Doss C (2011) The role of women in agriculture. ESA Working Paper No. 11-02. Agricultural Development Economics Division: The Food and Agriculture Organization of the United

Nations. Available via http://www.fao.org/3/am307e/am307e00.pdf. Accessed 16 Apr 2019

Steel G, Cottyn I, van Lindert P (2017) New connections—new dependencies: spatial and digital flows in Sub-Saharan African livelihoods. In: De Haan L (ed) Livelihoods and development: new perspectives. Brill, Leiden, pp 148–167

Steinbrink M (2007) Exploring the role of migrants' networks in the context of translocal livelihoods in South Africa. In: Lohnert B (ed) Social networks: potential and constraints. Indications from South Africa. Verlag für Entwicklungspolitik, Saarbrücken, pp 73–113

Steinbrink M (2009) Leben zwischen Land und Stadt, Migration, Translokalität und Verwundbarkeit in Südafrika. Springer, Wiesbaden

Sterly H (2015) 'Without a mobile phone i suppose i had to go there'—Mobile communication and translocal social constellations in Bangladesh. ASIEN 134(1):31–46

Sward J (2016) Moving to 'Greener Pastures'? The complex relationship between internal migration, land tenure and poverty in mid-Ghana. RPC Working Paper No. 33. Migrating out of Poverty Consortium, University of Sussex, Brighton

Tacoli C (2004) Rural-urban linkages and pro-poor agricultural growth: an overview. Paper prepared for the OECD DAC POVNET Helsinki workshop 17–18 June 2004

Tacoli C (2009) Crisis or adaptation? Migration and climate change in a context of high mobility. International Institute for Environmental and Development Studies (IIED). Environment & Urbanization 21 (2):513–525

Tacoli C (2011) Not only climate change: mobility, vulnerability and socio–economic transformations in environmentally fragile areas of Bolivia, Senegal, and Tanzania. Human Settlements Working Paper Series—Rural–Urban Interactions and Livelihood Strategies (28)

Tanser F, Bärnighausen T, Grapsa E, Zaidi J, Newell M-L (2013) High coverage of ART associated with decline in risk of HIV acquisition in rural KwaZulu-Natal, South Africa. Science 339(6122):966–971

Tanser F, Gijsbertsen B, Herbst K (2006) Modelling and understanding primary health care accessibility and utilization in rural South Africa: an exploration using a geographical information system. Soc Sci Med 63(3):691–705

Tiffen M, Mortimore M, Gichuki F (1994) More people, less erosion: environmental recovery in Kenya. Wiley, Chichester

Tompkins EL, Adger WN (2004) Does adaptive management of natural resources enhance resilience to climate change? Ecol Soc 9(2):10

Trager L (1995) Women migrants and rural-urban linkages in South-Western Nigeria. In: Baker J, Akin Aida T (eds) The migration experience in Africa. Nordiska Afrikainstitutet, Uppsala, pp 303–320

UN (2018) Economic development in Africa Report. Migration for structural transformation. New York, Geneva

UNAIDS (2014) The gap report. UNAIDS. http://files.unaids.org/en/media/unaids/contentassets/documents/unaidspublication/2014/UNAIDS_Gap_report_en.pdf. Accessed 18 Mar 2019

UNESCO (2012) Youth and skills: putting education to work. Available via UNESDOC. https://unesdoc.unesco.org/ark:/48223/pf0000218003. Accessed 18 Mar 2019

van der Land V, Hummel D (2013) Vulnerability and the role of education in environmentally induced migration in Mali and Senegal. Ecol Soc 18(4):14

Vargas-Lundius R, Lanly G, Villarreal M, Osorio M (2008) International migration, remittances and rural development. Available via FAO. http://www.fao.org/3/a-ak405e.pdf. Accessed 18 Mar 2019

Verspoor AM (2008) The challenge of learning: improving the quality of basic education in Sub-Saharan Africa. In: Johnson D (ed) The changing landscape of education in Africa: quality, equality and democracy. Symposium Books, Oxford, pp 13–43

Vorlaufer K (1985) Frauen-migration und Sozialer Wandel in Afrika. Das Beispiel Kenya. Erdkunde 39(2):128–143

Waddington H, Sabates-Wheeler R (2003) How does poverty affect migration choice? A review of literature. In: IDS Working Paper T3. Brighton

Warner K, Afifi T (2014) Where the rain falls: evidence from 8 countries on how vulnerable households use migration to manage the risk of rainfall variability and food insecurity. Clim Dev 6(1):1–17

Warner K, Hamza M, Oliver-Smith A, Renaud F, Julca A (2010) Climate change, environmental degradation and migration. Nat Hazards 55(3):689–715

Watkins K, Quattri M (2014) Lost in intermediation: how excessive charges undermine the benefits of remittances for Africa. Available via ODI, Overseas Development Institute. http://www.odi.org.uk/sites/odi.org.uk/files/odi-assets/publications-opinion-files/8901.pdf. Accessed 18 Mar 2019

Werthmann K (2007) Dans un monde masculin le travail de femmes dans un camp de chercheurs d'or au Burkina Faso. In: Boesen E, Marfaing L (eds) Les nouveaux urbains dans l'espace Sahara-Sahel. Karthala, Paris, pp 295–322

Wong M (2006) The gendered politics of remittances in Ghanaian transnational families. Econ Geogr 2(4):335–381

World Bank (2003) Global development finance 2003: striving for stability in development finance. 1: Analysis and statistical appendix. Available via The World Bank. http://documents.worldbank.org/curated/en/698051468128113998/pdf/multi0page.pdf. Accessed 18 Mar 2019

World Bank (2009) Remittances and natural disasters: ex-post response and contribution to ex-ante preparedness. Available via The World Bank. http://documents.worldbank.org/curated/en/951761468182368689/pdf/WPS4972.pdf. Accessed 18 Mar 2019

World Bank (2012) World Bank migration and remittances data. bilateral remittances matrix 2012, World Bank, World Development Indicators. Available via The World Bank. http://pubdocs.worldbank.org/pubdocs/publicdoc/2015/9/106901443117530403/Bilateral-Remittance-Matrix-2012.xlsx. Accessed 18 Mar 2019

World Bank (2013) Migration and development brief (Nr. 21) Migration and remittances team development prospects. Available via The World Bank. http://pubdocs.worldbank.org/en/471191444756853938/MigrationandDevelopmentBrief21.pdf. Accessed 18 Mar 2019

World Bank (2014) World Development Indicators 2014. Available via The World Bank. http://documents.worldbank.org/curated/en/752121468182353172/pdf/879460PUB0REPL00Box385214B00PUBLIC0.pdf. Accessed 22 Mar 2019

World Bank (2018) Migration and remittances. Recent developments and outlook. Special topic: transit migration. Available via World Bank Group. https://goo.gl/axrBaq. Accessed 18 Mar 2019

Wurster G (1995) Beruf und Karriere im Leben gebildeter Frauen in Nairobi, Kenia. Dissertation Thesis, Universität Bayreuth

Yaro JA, Awumbila M, Teye JK (2015) The life struggles and successes of the migrant construction worker in Accra, Ghana. Ghana J Geogr 7(2):113–131

Youngstedt SM (2013) Voluntary involuntary homebodies adaptations and lived experiences of Hausa 'left behind' in Niamey, Niger. In: Kane A, Leedy TH (eds) African migrations patterns and perspectives. Indiana University Press, Bloomington, Indianapolis, pp 133–157

Ziegelmayer U, Spaan E (2019) Migrant trajectories within the context of demographic, socio-economic and environmental change: evidence from coastal Ghana. In: Hillmann F, Van Naerssen T, Spaan E (eds) Trajectories and imaginaries in migration: the migrant actor in transnational space. Routledge, London, New York, pp 34–55

Zourkaleini Y, Mimche H, Nganawara D et al (2013) L'Impact des migrations Sud-Sud sur le développement au Cameroun (The impact of South–South Migration on Development in Cameroon). ACP

Zuma K, Gouws E, Williams B, Lurie M (2003) Risk factors for HIV infection among women in Carletonville, South Africa: migration, demography and sexually transmitted diseases. Int J STD AIDS 14 (12):814–817

Chapter 7
Translocal Development?

Abstract This chapter concludes this study by summarizing the results of the book and placing them in their larger context. We draw some general conclusions concerning the relationship between migration, translocality, and "(rural) development" and give answers to the following questions:

(1) Do translocal livelihoods have a balancing effect on structural inequality? (Sect. 7.1)
(2) Is the translocality of livelihoods in sub-Saharan Africa a temporary phenomenon? (Sect. 7.2)
(3) Is the translocality of livelihoods a problem of "(rural) development" or a solution for it? (Sect. 7.3)

At the end of this book, we advocate for a *translocal awareness* in development practice in order to make *translocal development* possible.

Keywords Inequality · Poverty · Translocal development

The preceding chapters have demonstrated that the phenomenon of translocal livelihoods is extremely prevalent in Africa south of the Sahara. Existing figures on the scope of circular migration and remittances, both international and domestic, indicate that a considerable proportion of the population in sub-Saharan Africa is integrated into translocal contexts (as movers or stayers)—according to the available studies, an estimated 40–60% of the population from rural regions and about 70% of the urban population, i.e., in total around half a billion people (see Sect. 5.2).

These numbers clearly demonstrate that the sedentarism and the local organization of everyday life are by no means the universal norm in Africa. Migration is an integral part of

M. Steinbrink and H. Niedenführ, *Africa on the Move*, Springer Geography,
https://doi.org/10.1007/978-3-030-22841-5_7

African livelihoods as they are the accompanying social networks that span over sometimes vast distances and across territorial boundaries. Accordingly, the life contexts in rural and urban areas are strongly interlinked, which is why we should consequently be speaking of *translocal* living conditions, instead of the differentiation of rural and urban living conditions.

To the extent that it overcomes its container-spatial bias, the livelihood and vulnerability perspective allows us to grasp different forms of economic, migratory, and social interaction between different (sometimes very distant) places as central aspects of livelihood security. The concept of translocal livelihoods (see Chap. 3) and the model of the translocal (see Chap. 4) are highly instructive for understanding (rural–urban or rural–rural) socio-spatial interlinkages in Africa.

Moreover, it makes sense to view the structure of socio-spatial translocality as the (intended and unintended) consequence of the social practices and rational actions of individuals and households. For that reason, at various points in this book, we have emphasized the importance of accounting for the context dependency of rational action (i.e., the variation in objective setting according to the degree of vulnerability) when investigating translocal livelihood strategies (migration, translocal economic diversification, translocal social networking, etc.). There are two reasons for this: First, it undermines any hasty equation of rational action with free action, instead of showing that the translocal livelihood structures in Africa are often based on action logic that is subject to strict context-related constraints. Second, distinguishing between the *survival, safety, and maximization rationalities* (Sect. 4.2) allows a more differentiated view of the *outcomes* of translocal action and its effects on processes of change/development (in rural areas). Thus, the theoretical approach itself conveys why the question of translocality's effects is by no means easily answered.

This book's analysis of case studies on (circular) migration, space-spanning networks, interaction, and exchange in sub-Saharan Africa reveals the extent to which the translocality of livelihoods is the result of actors handling their specific situations. The deteriorating living conditions in many African states—in rural and urban areas alike—call for us to understand translocal interlinkages above all as a product of vulnerability; for it is predominantly vulnerable population groups that resort to and rely on the strategy of translocal household organization. In this respect, translocal structures are, more often than not, the result of action aimed at coping with crisis and/or reducing risk and uncertainty.

However, vulnerability is not only a trigger for translocalization processes but also, simultaneously, a defining structural factor of the conditions for action: Permanent translocal socio-spatial structures evolve, and in this way, the context in which action is taken and to which livelihood systems must adapt becomes itself translocal. As such, vulnerability is not local, neither urban nor rural, but social.

Vulnerability in sub-Saharan Africa is often a translocal phenomenon.

In this sense, vulnerability in sub-Saharan Africa is often a translocal phenomenon. This means that, for many people, rural and urban vulnerabilities are just the two sides of *one* translocal plight.

Although, as many case studies show, the translocality of livelihoods can have a mitigating effect on vulnerability, as it expands the scope of action of those involved, the local situation in one locality becomes directly dependent upon that in the other interlinked locality. Both changes in the framework conditions influencing action in one place and the actions themselves can directly or indirectly impact actors in other places.

Translocal livelihoods result in a quasi-spatial fabric of interdependency in which local influences have translocal effects.

Consequently, translocal livelihoods result in a quasi-spatial fabric of interdependency in which local influences have translocal effects. Accordingly, a "ruralization of urban crises" or an "urbanization of rural risks" are entirely feasible in this context of rural–urban socio-spatial interconnectedness.

Therefore, the question arises: What is the relationship between translocality and rural structural change?

If one understands "rural structural change"—as proposed by Rauch et al. (2016: iv)—as a "long-term multidimensional process changing the fundamental characteristics of the economics and lifestyle of people in rural areas, in light of their embeddedness in macrosocial and global dynamics [transl.]," then one must regard the process of translocalization and the translocality of livelihoods as an essential expression of this structural change. In the course of livelihood translocalization, economic and lifestyle patterns are changing dramatically, primarily through the transformation of socio-spatial structures, which are expanding and increasingly spanning rural–urban boundaries. This expansion affects almost every aspect of the lives of those operating in such translocal fields—economic, social, and cultural. Can this far-reaching structural change then still accurately be called *rural* structural change?

Nonetheless, at least some general conclusions concerning the relationship between translocality and "(rural) development" can be drawn. Therefore, the final focus lies on the following questions:

(1) Do translocal livelihoods have a balancing effect on the (spatial) pattern of structural inequality? (see Sect. 7.1)
(2) Is the translocality of livelihoods in sub-Saharan Africa a temporary phenomenon? (see Sect. 7.2)
(3) Is the translocality of livelihoods a problem of "(rural) development" or a solution for it? (see Sect. 7.3)

7.1 Translocal Livelihoods and the Persistence of Inequality

Structural inequalities between regions and between urban and rural areas of a country are generally interpreted as development problems, since they are associated with socioeconomic disparities, unequal distribution of macroeconomic returns, unequal living conditions, and thus injustice.

Translocal livelihoods only develop because of structural inequalities.

Ultimately, translocal livelihoods only develop based on existing structural inequalities. That is because the translocal organization of livelihood is always a matter of combining different economic activities and using the different opportunities and infrastructure at different locations, while striving to minimize the cost of living. Translocal livelihood organization is thus about fusing, by means of social networks, the opportunity structures existing at separate places into a single opportunity structure in a translocal social space. To put it pointedly, if there were no spatial pattern of structural inequality, there would be no translocal livelihoods.

As plausible and correct as it is then to understand spatial-structural inequalities as a trigger for migration and translocalization processes, it would be rash to infer from this a balancing tendency in the sense of neoclassical equilibrium models. Migration and translocalization may even contribute to the reproduction of this pattern of disparities.

An initial look at the remittances conveyed in translocal flows (Sect. 6.1) shows that a substantial part of the income earned by migrants in cities or structurally stronger regions is transferred to their rural areas of origin. Thus, there seems to be a powerful flow of capital from economically stronger, prosperous regions to structurally weaker, poorer ones. If we consider the socioeconomic conditions of the vast majority of migrant workers, however, what at first glance appears to be a large regional compensatory current often turns out to be merely a redistribution between "the poor here and the poor there." Thus, translocal structures should not be perceived as a medium of social redistribution between poor and rich; instead, the vulnerable pay for the vulnerable. Furthermore, this transfer takes place within translocal household structures, meaning it is part of those internal distribution processes that do not normally interest macroeconomists. These payments alone do not make households wealthier!

To answer the question of whether migrant remittances nonetheless positively affect rural development, leading to the reduction of spatial-structural or social disparities in the long term, it is helpful to look at the *use* of these transfer payments.

Remittances are of immense importance for securing the livelihood of the rural population.

That remittances are of immense importance for securing the livelihood of the rural residential population was demonstrated in Sect. 6.1.1. After all, an estimate of 20–40% of households in rural sub-Saharan Africa regularly receives remittances, and about 20% of the monetary income available in rural areas comes from remittances. However, the vast majority of remittances are not invested in *productive assets* but applied by recipients in rural areas directly toward securing or improving living standards (Sect. 6.1.1.3). In other words, the funds are primarily used for consumption purposes, with only a small portion being invested toward entrepreneurial ends.

Productive investments in the agricultural or nonagricultural sector are made—if at all—by households that are already better off. The majority of translocal households operating at the subsistence level cannot afford productive investment, either because the available financial resources are insufficient or because the risk of a misinvestment is considered too great.

Remittances hardly stimulate any endogenous economic dynamics in local rural economies.

In this respect, we can assume that remittances hardly stimulate any endogenous growth dynamics in local rural economies. And although there are examples of remittances, in particular from international migrants, having positive economic effects in their regions of origin, the rule holds true: As long as a large proportion of rural households are operating in the context of high vulnerability, remittances cannot be expected to produce "economic miracles" in rural sub-Saharan Africa.

The translocality of livelihoods often creates a growth-inhibiting anomaly in the demographic structure in rural areas.

Moreover, the translocality of livelihoods often creates a growth-inhibiting anomaly in the demographic structure of the rural residential population, which makes autonomous rural economic development much more difficult. A large part of the working population leaves rural areas, and it is often the young and more innovative who hope for better prospects in urban centers (see Sect. 5.3.4). The effort to compensate for the lack of family workforce in rural regions is a heavy burden on those left behind, who must shoulder multiple roles. The "feminization of agriculture" observed in this context (see Sect. 6.3.2.1) has led to a further deterioration in production conditions and the market position of small-scale farms in many places, not least because of the persistence of the weaker social positioning of women in Africa and their limited decision-making power.

All of this contributes to the fact that the most important resource in the countryside—the land—is often used ineffectively or even withdrawn completely from agricultural use. This is one reason why, under present conditions, the growth opportunities in the agricultural sector afforded by global market trends often remain unexploited. This also explains why the translocality of livelihoods and income transfers do not help to equalize regional disparities.

Spatial disparities are exacerbated by the system of translocal livelihoods.

From a political-economic point of view, it could even be argued that spatial disparities are exacerbated by the system of translocal livelihoods: The difference between rural and urban living costs is one important reason for translocal household organization. The cost of living in migrants' rural areas of origin is usually much lower. Reasons for this include the contribution of subsistence farming, the absence of land and housing markets, and the provision of supplies, like firewood and building materials, outside of monetary cycles. This "comparative cost advantage" is utilized as some members of the translocal household—especially those who do not or cannot pursue paid employment—remain in rural areas. However, because of the configuration of the labor market in African cities, the savings do not primarily benefit laborers' households. Instead, the urban labor market situation reinforces the economic pressure on many jobseekers to accept even extremely low earnings. Because of the oversupply of (low-skilled) labor, wages in the cities are usually so low that they hardly guarantee the subsistence level of the workers and their families. The savings realized through the translocal organization of livelihood systems thus act as an informal subsidization of labor costs. Part of the reproduction costs for the work carried out in urban areas is externalized to the rural (subsistence) sector. This means that the savings generated by rural–urban cost differentials flow into the added value, which is skimmed off as profit by people who can hardly be considered vulnerable.

Translocal livelihoods in Africa will likely continue to reinforce existing patterns of spatial disparity and social inequality.

In this respect, the system of translocal livelihoods in Africa will likely continue to exacerbate and reinforce existing patterns of spatial disparity and social inequality in the future.

7.2 Translocal Livelihoods as a Temporally Stable Phenomenon

The assumption that the translocality of livelihoods is only a transitive phenomenon appears unlikely.

If we deduce that spatial-structural disparities are causal for translocalization processes, and if at the same time we find that the translocality of livelihoods does not reduce inequalities (see Sect. 7.1), then the assumption that the translocality of livelihoods is only a transitive phenomenon already appears unlikely on a logical level. And after reviewing the available empirical case studies, we see little reason to assume that translocal systems of livelihood organization will lose importance in the near future or represent merely a transitional phenomenon in the African urbanization process, particularly since recent literature has widely indicated that space-spanning forms of livelihood organization are still gaining ground. Here, mobile communication technologies and the internet play a decisive role, because they facilitate and strengthen translocal network structures, albeit at high financial costs to translocal households (see Sects. 5.3.5 and 6.1.1.1).

As long as translocal livelihoods make sense, they will persist.

The translocalization of livelihoods should be understood as the result of adaptation to changing conditions and options for action and, beyond that, as the result of rational (household) action. This means that as long as being embedded in translocal life contexts seems expedient to the actors involved, such translocal life contexts will continue to exist.

The translocalization of livelihoods is not so much a question of desire as of necessity.

Many of the studies at hand can be interpreted to the effect that translocal action (transmigration, translocal income diversification, and translocal social networking) is geared less toward maximization than toward safety and risk avoidance. This is demonstrated by the high proportion of vulnerable and economically weak households who organize their lives in translocal contexts. The main cause for the emergence and persistence of translocal interconnectedness in Africa is not the great opportunities for prosperity associated with it, but rather the lack of alternatives: In sum, for the majority of households, the translocalization of livelihoods is not so much a question of desire as of necessity.

The driving forces behind translocalization processes can by no means be clearly localized in either rural or urban areas, because on both sides (structural) conditions have a causal effect on translocality. The situation of agricultural production (including susceptibility to crisis due to market developments and climatic uncertainties), the difficulties in market integration as well as the insufficient nonagricultural employment opportunities in rural areas make it largely impossible for many people to secure a "sedentary livelihood" in the countryside, which in turn makes strategic migration unavoidable. Extremely low wages and the various uncertainties of the urban labor market, poor and unsafe housing conditions, and comparatively high living costs, on the other hand, make the migration of all household members impossible. Therefore, many households are forced to combine different local opportunities, which alone would not guarantee a livelihood, translocally.

Ultimately, translocal structures evolve and link various local vulnerability constellations. This is a result of action aimed at mitigating the respective risks. However, the risk factors themselves remain unchanged; the risks are merely distributed (socially, economically, and spatially) in order to moderate their potential impact.

Socio-spatial interconnectedness is stabilized by the permanent instability of living conditions in the socially interlinked localities.

Many of the aspects addressed in this book suggest that vulnerability not only contributes to the emergence of translocal structures but also to their reinforcement. In other words, spatial interconnectedness is stabilized by the permanent instability of living conditions in the socially interlinked localities. An evaluation of the available studies leads us to conclude that the translocal organization of households is indeed a suitable strategy for dealing with respective rural and urban insecurities; however, the translocality of livelihoods alone is no means for reducing the necessity of translocality. For this, translocality too rarely entails a substantial improvement in living conditions in the interlinked places.

The necessity of translocality persists in both rural and urban contexts, and it arises to a large extent from the translocal social realities themselves.

Nevertheless, the statement that the situation would be even worse *without* translocalization remains hypothetical, if plausible.

Nota bene: The necessity of translocality persists in both rural and urban contexts, and it arises to a large extent from the translocal social realities themselves.

7.3 Plea for a *Translocal Awareness* in Development Planning

Should—in light of the problems of poverty and vulnerability in rural sub-Saharan Africa—the translocality of livelihood security then be positively or negatively appraised?

Translocal livelihoods are both solution and problem, and also neither.

Many readers, especially development practitioners, may have hoped for an answer to this question *before* reading this book. For our part, we now hope that, *after* reading it, it is clear why there can be no definitive answer. Ultimately, the issue is not whether translocal livelihoods are a solution or problem for rural development in Africa; they are both and also neither. But even this ambivalent insight has great practical relevance: It shows that development policy and development cooperation should neither develop specific measures to "promote translocality" nor take direct "measures against translocality."

From the correct observation that the translocality of livelihoods results from dynamic adaptation processes, one ought not to simply conclude that the result is an "adapted solution" worth supporting as a self-help effort of the target group. Such a conclusion would ignore the fact that translocality, by its very nature, is often an expression of constraint and lack of alternatives, and that it may perpetuate the actors' vulnerability. On the other hand, if developmental practitioners focused on the important realization that translocality is often both a manifestation and a driver of vulnerability and if they thus interpret the translocality of livelihoods as a problem to be solved, they would fail to recognize the existential necessity of this space-spanning form of household organization.

At the end of this book, we thus recommend that development practitioners not focus primarily on measures aimed at the translocal systems themselves. Any such measures are best limited to minimizing translocal "frictional losses" arising from the costs and effort of transport, communication, and transfer (e.g., of remittances).

Translocality is the context of practical development work.

Apart from that, what is most important in practice is recognizing *translocality as the context of practical development work* and accepting translocal livelihoods as a fact, understanding their dynamics and taking them into account. Planners should replace container-spatial (rural/urban) thinking with the insight that development measures that affect translocally organized target groups always work within their translocal life contexts.

Such a *translocal awareness in development planning* means, for one, acknowledging that locally implemented measures almost inevitably have space-spanning effects such that intervention on one side also has repercussions on the other side of the translocal system, as well as on the translocal system itself. Think, for example, of the impact of land reforms: A transfer of communal land under traditional land use rights into private ownership could profoundly disrupt translocal systems of livelihood security. Such a step would possibly necessitate a complete reorganization of the livelihoods with all attendant risks for the affected households and far-reaching consequences, e.g., for urbanization and urban development in Africa.

Conceiving translocal structures as a necessary condition for the successful implementation of projects and measures.

On the other hand, a *translocal awareness in development planning* also means conceiving of translocal structures as a necessary condition for the successful implementation of projects and measures. This is exemplified by special programs aimed at providing women with farming advice and agricultural extension services: Such programs would easily be doomed to failure if the male (*de jure*) heads of household residing elsewhere were not involved in the process. For, despite their absence, their involvement in the agricultural decisions of the household is often decisive.

Complex translocal interdependencies undoubtedly place considerable demands on development planning, but the translocal planning perspective is indispensable when seeking to do justice to the social realities in rural sub-Saharan Africa. In this respect, this book primarily serves as a plea for a (spatially) integrating view of "rural development challenges" in Africa.

Reducing the "need for translocality."

There are no patent recipes for planning. In general, however, all those measures that contribute to reducing the "need for translocality" can be considered reasonable, irrespective of whether they are implemented in rural or urban areas. Through positive changes in local vulnerability contexts—be it in the city or in the countryside—the actors gain leeway and freedom of action.[1] While we cannot say whether the actors involved would use this additional freedom of action for relocalization and sedentariness (here or there, in a rural area or a city) or for the establishment of new structures and forms of translocality, which could offer greater opportunities for prosperity and well-being, it would be "better" in any case.

Make translocal development possible!

And when the translocality of livelihoods ultimately no longer expresses an absence of alternatives but develops into a chance for people to effectively improve their own situation in life, then *translocal development* will become possible.

References

Rauch T, Beckmann G, Neubert S, Rettberg S (2016) Ländlicher Strukturwandel in Subsahara Africa. SLE Discussion Paper 01/2016. Available via SLE. https://www.sle-berlin.de/files/SLE%20Discussion%20Paper%2001-Strukturwandel%20mit%20Cover.pdf. Accessed 18 Mar 2019

UN (2018) Economic development in Africa Report. Migration for Structural Transformation. New York, Geneva

[1]As the UN (2018: 4) states: "Migration should always be a choice and not a necessity." This holds true for translocality, too.

Index

M. Steinbrink and H. Niedenführ, *Africa on the Move*, Springer Geography,
https://doi.org/10.1007/978-3-030-22841-5

MIX
Papier aus verantwortungsvollen Quellen
Paper from responsible sources
FSC® C105338

If you have any concerns about our products,
you can contact us on
ProductSafety@springernature.com

In case Publisher is established outside the EU,
the EU authorized representative is:
Springer Nature Customer Service Center GmbH
Europaplatz 3, 69115 Heidelberg, Germany

Printed by Libri Plureos GmbH
in Hamburg, Germany